ENTANGLEMENTS OF THE MAYA UNIVERSE

Ethnography, Theory, Experiment

In recent years, ethnography has been increasingly recognized as a core method for generating qualitative data within the social sciences and humanities. This series explores a more radical, methodological potential of ethnography: its role as an arena of theoretical experimentation. It includes volumes that call for a rethinking of the relationship between ethnography and theory in order to question, and experimentally transform, existing understandings of the contemporary world.

Recent volumes:

Volume 11
ENTANGLEMENTS OF THE MAYA UNIVERSE
LIFE, SOCIETY AND COSMOS AMONG THE K'ICHE' OF MOMOSTENANGO
By Alonso Zamora Corona

Volume 10
DREAMING AND THE IMAGINATION
THEORETICAL INTERSECTIONS IN CROSS-CULTURAL PERSPECTIVE
Edited by Matthew D. Newsom

Volume 9
CATTLE POETICS
HOW AESTHETICS SHAPES POLITICS IN MURSILAND, ETHIOPIA
By Jean-Baptiste Eczet

Volume 8
THE CHILDREN OF GREGORIA
DOGME ETHNOGRAPHY OF A MEXICAN FAMILY
By Regnar Kristensen and Claudia Adeath Villamil

Volume 7
GOING TO PENTECOST
AN EXPERIMENTAL APPROACH TO STUDIES IN PENTECOSTALISM
By Annelin Eriksen, Ruy Llera Blanes and Michelle MacCarthy

Volume 6
CUTTING COSMOS
MASCULINITY AND SPECTACULAR EVENTS AMONG THE BUGKALOT
By Henrik Hvenegaard Mikkelsen

Volume 5
VITAL DIPLOMACY
THE RITUAL EVERYDAY ON A DAMMED RIVER IN AMAZONIA
By Chloe Nahum-Claudel

Volume 4
VIOLENT BECOMINGS
STATE FORMATION, SOCIALITY, AND POWER IN MOZAMBIQUE
By Bjørn Enge Bertelsen

Volume 3
WATERWORLDS
ANTHROPOLOGY IN FLUID ENVIRONMENTS
Edited by Kirsten Hastrup and Frida Hastrup

Volume 2
FIGURATIONS OF THE FUTURE
FORMS AND TEMPORALITIES OF LEFT RADICAL POLITICS IN NORTHERN EUROPE
By Stine Krøijer

For a full volume listing, please see the series page on our website: https://www.berghahnbooks.com/series/ethnography

ENTANGLEMENTS OF THE MAYA UNIVERSE
Life, Society and Cosmos among the K'iche' of Momostenango

Alonso Zamora Corona

berghahn
NEW YORK • OXFORD
www.berghahnbooks.com

First published in 2025 by
Berghahn Books
www.berghahnbooks.com

Library of Congress Cataloging-in-Publication Data

A catalogue record for this book is available from the Library of Congress

British Library Cataloguing in Publication Data

A catalogue record for this book is available from the British Library

EU GPSR Authorized Representative

LOGOS EUROPE, 9 rue Nicolas Poussin, 17000, LA ROCHELLE, France
Email: Contact@logoseurope.eu

ISBN 978-1-83695-026-4 hardback
ISBN 978-1-83695-027-1 epub
ISBN 978-1-83695-028-8 web pdf

https://doi.org/10.3167/9781836950264

To my late parents and my brother

We the gods have arrived, we who are gathered,
The Earth gods, the lords of the face of the world.

—K'iche' prayer

CONTENTS

ILLUSTRATIONS

Figures

Maps

ACKNOWLEDGEMENTS

The publication of this book is the end of a long journey in which I have become indebted to the generosity of many people. Before naming the people to whom I owe my gratitude, from Mexico to London (and back again), I would like to thank the people of Momostenango, especially those present in this ethnography, for without their help, hospitality and protection, I would never have finished this book. To each and every one of them there I want to say: *ix wajch'ab'e'l chbe q'iij chbe saq*. You know that these words are written now in the books of the ancestors, and, as you insisted to me since the first day, they will never be erased from 'the face of the world'.

This book is the result of sixteen months of fieldwork, completed in October 2018, as part of a PhD project funded both by the Consejo Nacional de Humanidades, Ciencias y Tecnologías (CONAHCYT) in Mexico, and by University College London, in the United Kingdom. Among my colleagues in social anthropology and ethnology, I would like to thank, in no particular order: Gabriela Rivera, Fernando Guerrero, Stan Declercq, Daniel Moreno Zaragoza, Sanja Savkić, Carlos Mondragón, Sergio González Varela, Gabriel Kruell, Johannes Neurath, Natalia Gabayet and Alessandro Questa, all of whom have been generous friends and interlocutors of this ethnography. Particularly, I am grateful to Canek Estrada Peña, who introduced me to Momostenango.

Among those whom I worked with in London, I would like to thank, first and foremost, Martin Holbraad, for his guidance and friendship, and Lucia Michelutti, for her encouragement and help. Also, in a vague semblance of a chronological order, I would like to thank some of the friends I met in Europe, which made me feel more like home in a continent in which I was a stranger: Keiko Homewood, Sabine de Graaf, Miriam Lamrani, Tobia Farnetti, Hilary Leathem and Łukasz Byrski.

Finally, returning to Mexico, I would like to thank my long-time friends: Ricardo Pérez Martínez, Mariana Ozuna Castañeda, Alicia Mastretta Yanes, Carlos Gallardo Soto, Laura Angélica de la Torre, Jorge Arellano Olvera, Artemio Urbina, Iván Alanis Ortiz, Megan Thomas, Pablo Yanes Thomas and Eduardo Medina Frías, all who were generous and supportive of my work and my person during the many vicissitudes of life that accompanied this project. Finally, I would like to thank my sister, Sol de María Zamora Corona, without whose love and constant support this book would never have been written. This book is dedicated to the memory of my late parents, María Teresa Corona Agúndez and David Zamora Ramírez, and to my late brother, Julio David Zamora Corona.

NOTE ON K'ICHE' ORTHOGRAPHY AND TRANSLITERATIONS

The K'iche' orthography used in this book is that established by the Academy of Maya Languages of Guatemala (Cerezo Arévalo 1987). All the translations from Spanish are mine, and all the translations from K'iche' are by Abraham Ajiataz.

Map 0.1. Momostenango within Guatemala. © Alonso Zamora Corona

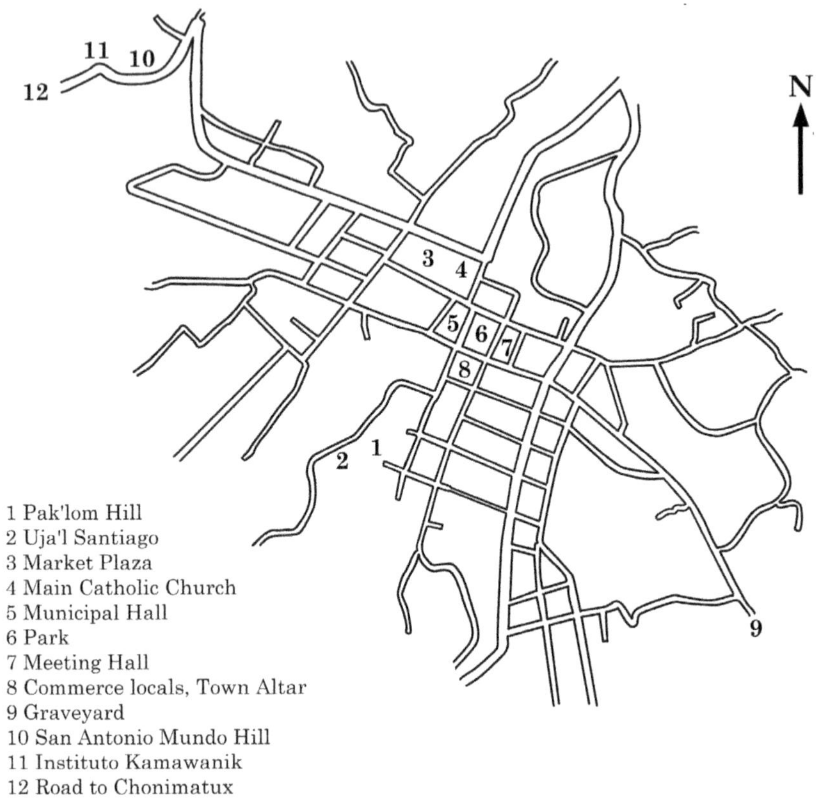

Map 0.2. Momostenango's town centre and some places of interest to my ethnography. © Alonso Zamora Corona

Introduction

Mesoamerican Cosmologies and Social Theory

Cosmology, which has been defined as 'the theory of the cosmos as an ordered whole' (Howell 2012: 157), is considered a classic topic both in anthropology in general and in the anthropology of Indigenous groups of 'Mesoamerican' tradition both past and present,[1] because of their complex and not wholly understood ideas regarding notions such as space, time, personhood and the relationship of those notions with social life, history and power. In recent years, however, cosmology as a topic in social anthropology has been viewed with an increasing 'weariness' (Abramson and Holbraad 2014: 2), and has entered what could be considered a deep crisis, a crisis that could probably mean none other than the definitive decline of the topic in the discipline; such a crisis, of course, has also impacted the debate on Mesoamerican cosmologies, with an increasing number of scholars calling us to rethink established assumptions of all kinds (Díaz 2020). Thus, despite what the title of this book may suggest, the purpose of this ethnography is not to offer yet another ethnological depiction of the cosmology of a particular group, the K'iche' Maya of Momostenango, Guatemala, a topic that has been the subject of a number of classic ethnographies.[2] Rather, the purpose of this book is to address the problem of cosmology in anthropology from the perspective of my ethnographic encounter with the many people that this book will endeavour to portray. The important question of this book is not *how* the cosmology of the K'iche' is, even if in some of its pages the reader will indeed find a summary of the topic and even new insights and information about it; rather, the questions of this book are:

How does this cosmology and other cosmologies *work*? Why do people enter into contact with them or 'turn to them' in order to solve problems in their lives? Why are cosmologies in general 'efficient'? And, finally: How can we reconceptualize or redeploy the concept of cosmology in anthropology itself? Is there a possible 'shape' behind 'non-Western' cosmological thinking? However, before I start my ethnographic account and introduce the possible new theoretical avenues suggested by it, it is necessary to contextualize and examine two things more deeply: the current crisis of cosmology in anthropology as such, and the debate regarding Indigenous cosmologies in Mesoamerican anthropology, with a particular emphasis on the Guatemalan context.

The Cosmological Crisis in Anthropology

As stated by Allen Abramson and Martin Holbraad in the prologue of a collective volume dedicated to the topic (2014), the crisis of cosmology in anthropology is both the crisis of its relevance within the anthropology of the contemporary world, and the crisis of its conceptualization within classical anthropological theories. Regarding its relevance, in recent years many scholars have questioned that the sort of unified, totalizing, closed systems of thought that classic anthropologists used to depict as characteristic of non-Western groups, as well as the social isomorphisms supposedly necessary to their working, are even extant nowadays, given the consequences of colonialism and urbanization in the developing world, which have fragmented both our cosmos and that of the 'others' of anthropology (Abramson and Holbraad 2014: 1–2). In the case of Guatemala, for example, the old paradigm that considered Indigenous people as living in 'closed corporate peasant communities' (Wolf 1957) has been continuously discussed, redefined or redeployed in modern ethnographies due to ever increasing social changes that have challenged that paradigm in various ways (Brintnall 1979; Carlsen 2011; MacKenzie 2016; Watanabe 1992; Wilson 1995), changes that naturally have had an impact on how the cosmologies of Indigenous societies have been portrayed in recent years.

Deeply linked to this historical crisis is the crisis of the classic ways in which anthropology depicted Indigenous cosmologies throughout the world. Indeed, in the aforementioned piece, both Abramson and Holbraad present a series of criticisms to the notions that underlie the treatment of cosmology as a subject in classical social anthropology, notions that were problematic to begin with, and that have conditioned the recent decline of the subject in the discipline. Thus, the holistic, static, totalizing and organic cosmologies depicted in classical anthropology were somewhat a reflection of the hierarchical, ethnocentric way in which anthropological theory was constructed during the twentieth century, a theory that could be characterized itself as 'anthropology's cosmology of the social' (Abramson and Holbraad 2014: 7). In general, during the twentieth century, anthropological theories of all kinds tended to coincide in the notion that 'Indigenous cosmologies' were more or less holistic totalities that could be mapped into the social, in the assessment that they were scientifically *false*, and in that they were mere reflections of the more important and 'more rational' underlying principles uncovered by the theoretical activity of social scientists:

> ... basic human needs (functionalism), moral and socio-political order and reproduction (structural-functionalism), ecological adaptation (cultural materialism), individual agency (methodological individualism), the expression of underlying social values (interpretivism, symbolism), situated social relations (practice theory) or of a gestalt personality (culture and personality school), ideology and false consciousness (Marxism): all of these classic anthropological positions were posited as competing explanations as to why societies the world over set such great store in imagining the totality of the world in ways that have to be recognised as false. (Abramson and Holbraad 2014: 8)

Thus, what anthropologists traditionally sought to do was to try to subordinate Indigenous cosmologies to their own cosmologies of the social, while conceptualizing them as false, ideological or 'primitive'. However, as soon as these classical anthropological perspectives entered in a succession of crises and the pre-globalization conditions from which they arose started to fade, the topic of cosmology went into a continuous decline, from which perhaps it has not really escaped.

Naturally, the aforementioned succession of classical anthropological theories and their nowadays increasingly questioned assumptions are to be found in the history of the ethnology of Indigenous Maya communities in Mexico and Guatemala. Thus, the initial ideas of Robert Redfield and his collaborators regarding the 'little community', its social structures and practices, and what he called its 'primitive worldview' (Adams 1952; Guiteras Holmes 1961; Redfield 1941, 1952; Tax 1941), ideas partially inspired by the work of Marcel Griaule (1948), eventually led to the establishment of a functional-structuralist school that considered native cosmologies mostly as means of social control (Hermitte 1970; Holland 1961, 1964; Villa Rojas 1963, 1978). In turn, this functionalist school was consolidated by the monumental work of the Harvard Chiapas Project led by Evon Vogt and his collaborators (1969), which collected an impressive amount of information regarding such cosmologies and used it to create a normative model that sought to find isomorphisms between said Indigenous cosmologies and social hierarchies (Cancian 1976; Korsbaek 1990, 2017), as well as the archaeological past (Vogt 1964). Similarly, and despite Tax's own pioneering economic-anthropological study of Indigenous forms of capitalism (1963), Eric Wolf's ideas on 'closed corporate peasant communities' (1957) were generally accepted at that time as the dominant paradigm that described Indigenous societies, past and present. Similar models were endorsed by functionaries of the *indigenista* political movement in Mexico (Caso 1958) and Guatemala (Gobaud Carrera 1964), which sought the cultural assimilation of Indigenous communities to national states, and that considered Indigenous people as living in 'refuge zones', buffer zones that were more or less isolated from the influence of Western society, which had to partake of an inevitable 'acculturation process' facilitated and even enforced by the state (Aguirre Beltrán 1967).

Continuing earlier trends and applying them to the whole of Guatemalan society, with the works of Richard Adams (1952, 1968), a strong and even more decisive presence of American anthropology began to develop in Guatemala. This has been called 'the anthropology of occupation' by Guatemalan anthropologists and, in general lines, it developed along an 'ahistorical, structural functionalist' line (Pérez de Lara 1993: 19). Due to the deep political crises of Guatemala, which eventually led to the prolonged 'internal armed conflict' or civil war (1960–1996), the discipline entered into a crisis; thus, 'until 1974 the formation of human resources in anthropology was practically nonextant at the University of San Carlos', the most important of the country (Pérez de Lara 1993: 20). Thus, a void in Guatemalan local ethnography began, which only ended in the mid-eighties.

In Mexico, among a decade of political turmoil during the seventies, many critiques were levelled against the ahistorical, static assumptions of the functionalist school, the most prominent being those issued by Marxist anthropologists, who noticed that Indigenous communities were never really closed and were always in many different kinds of relationship with other communities, with Indigenous and colonial empires, and later with the national state (Albores Zárate 1978; Pozas Arciniegas 1974; Warman et al. 1969), a critique later endorsed by ethnohistorians (Viqueira Albán 2002), and even by former functionalists, given the dramatic changes undergone by the societies they studied in their youth (Cancian 1994). Similarly, later anthropologists working in Guatemala and other Indigenous communities of Mesoamerican tradition found that 'the cultural patterns described in … classic anthropological works

[were] no longer typical' (Brintnall 1979: 1), that modern Indigenous life was full of 'empty centres' (Watanabe 1992: 225), that communities underwent complex processes of ethnic reconstitution and dramatic confrontations with colonialism and the state (Wilson 1995), that 'closed corporate communities' had 'opened' (Carlsen 2011: 123), or had become transnational (MacKenzie 2016). On the other hand, politically, in a parallel manner to what happened in the rest of the world (cfr. Banaji 1970), the complicity between certain functionalist anthropologists with colonialism or, in the specific case of Latin America, with the policies of assimilation to national states, determined what was called in Mexico the 'political bankruptcy' of the theory (García Mora and Medina 1986), a bankruptcy that was ultimately symbolized by the seizing of the *indigenista* Coordinating Centre in San Cristóbal de las Casas, Chiapas, by the Zapatista National Liberation Army in 1994 (Lewis 2018: 1).

In Mexico, these crises, both theorical and political, paved the way for the emergence of the model that would dominate the latter part of the twentieth century in local ethnology, the historicist-symbolic model of Alfredo López Austin (1988). This model, called *teoría de la cosmovisión* (literally 'world view theory'), harked back to the earlier efforts of Redfield and Guiteras on the field of worldview studies and synthesized them with the historical ideas of the French Annals school, especially Braudel's conception of *longue durée* (1958), applied to the continuity of Mesoamerican conceptions, while asserting the ultimately 'ideological' character of Indigenous ideas.[3] In recent years, however, the model found itself in a crisis, brought by the onset of postmodernist, Deleuzian approaches to classical cosmological topics such as personhood (Pitarch 2010, 2013), critical calls to re-examine the traditional, essentialist assumptions of past anthropology (Neurath 2008) and, finally, the introduction of ideas of the so-called ontological turn in Mesoamerican ethnology (Lamrani 2008; MacKenzie 2016; Millán 2015, 2019; Zamora Corona 2020), as well the emergence of new contemporary orientations within the discipline such as cosmopolitics and the ecological crises of the present (Questa Rebolledo 2017, 2023).

However, despite the fact that the ontological turn has called anthropologists to take the ideas of the others seriously, and despite the fact that it has explicitly called to reject ethnocentrism, essentialism, exoticization and top-down explanations (cfr. Holbraad and Pedersen 2017: 5–6), a certain sense of continuity of the 'cosmological crisis' pervades the anthropology of Indigenous communities of Mesoamerican tradition, even after the arrival of the ontological turn, which, while asserting itself as more of a particular and more of a heuristic methodology than a repertoire of paradigmatic concepts that seeks to explain 'reality as such', somewhat ended up creating a set of concepts resembling such a repertoire. Indeed, for example, certain key topics introduced by its proponents, such as predation (Viveiros de Castro 1992), perspectivism (Viveiros de Castro 1998), affinity (Viveiros de Castro 2018), reciprocity (Descola 1986) and the different instances of the relationship between nature and culture (Descola 2013), to name a few, while of undeniable importance, have started to become more and more akin to common places among some of its followers. In this sense, a certain dissatisfaction with the history and current state of 'Mesoamerican' anthropology has been recently expressed by Maya ethnologist Pedro Pitarch in a recent volume dedicated to theory and ethnography in Mesoamerica, in which it is noted that the complexity of the ideas of Mesoamerican peoples 'has not found its place within the anthropological repertoire of the peoples of the world' (2020: 10), contrasting it with the rich conceptual tradition found within the anthropology of Melanesia and the Amazon. Thus, while the aforementioned volume suggests new avenues of enquiry and conceptualization, the overall impression is that a new impasse has been reached, an impasse whose causes have been addressed by those who are critical *within* the ontological turn, rather than merely *against* it.

As the Mexican anthropologist Carlos Mondragón, a specialist in Melanesia, remarked in a critical intervention at the Institute of Anthropological Research of the National Autonomous University of Mexico (2015), the danger of a sort of 'Kuhnian' substitution of a paradigm for another has been present in the reception of the ontological turn in Mexico, as well as a certain disdain for ethnography. Thus, for him, new engagements with ethnography within the ontological turn are needed, rather than the enthusiastic adoption of pre-existing topics; similarly, a more pronounced emphasis in processes is needed; that is, to study 'the daily conditions of reality that make possible the creation of the world at a local level'. To his very pertinent observations, I would add that the point of the ontological turn is not to 'verify' a 'new paradigm' by merely 'mapping' its emerging models into ethnographic data, nor to seek new ways to characterize 'Mesoamerican Indigenous peoples', but rather to attempt to engage with anthropological problems through the possibilities of an ontological approach in conjunction with new ethnography. Thus, this book ascribes itself to the ontological turn but will not pay that much attention to its now 'traditional' topics (which, nonetheless, are easy to identify in my ethnography), but rather will use the ontological outlook in order to address a more general, anthropological problem, that of cosmology itself.

In this sense, as a chart or guideline for future fructiferous engagements, Abramson and Holbraad have suggested that we should be wary of holistic schemes and totalizing pretentions (2014: 3), and have insisted that cosmology should stop being assumed as a 'baseline' for a cultural kind of explanation of people's lives, nor serve as a departure for totalizing and exotic discourses; instead, cosmology should be engaged with a spirit of ethnographical experimentation and imbrication with other dimensions of people's lives, taking the ethnographic contingency seriously (2014: 15). However, one wonders: if we are critical of the totalizing ideas of old-school social anthropology, and even more of the idea of totality itself, is cosmology still valid as a topic? Is *κόσμος* (and its modern calques along with it) not a word that conveys the idea of order and even totality?[4] If this has become problematic, why not abandon the term at all? Is cosmology not the activity of 'imagining the totality of the world' (Abramson and Holbraad 2014: 8)? Why not disperse, then, *cosmology* into some of its sub-topics like personhood, temporality or spirits, to name a few? Is it possible to even study cosmology by itself in a way that does not fall into the reductionist conundrums of former theory? How to talk about the totality par excellence, the *cosmos*, in a non-totalizing way? However, before assailing this topic, it is necessary to make some further clarifications in order to position this book in regards to some specific aspects of the ethnological tradition of Guatemala, both past and present.

Some Critical Clarifications

Besides dealing with cosmology in general, this book deals, in particular, with some of the cosmological ideas found in the K'iche' Maya town of Momostenango or Chwa Tzak, located in Totonicapán, in the Highlands of Guatemala. Perhaps some readers will notice that this book will distance itself from certain salient topics of the ethnographic tradition of the Highlands of Chiapas and Guatemala, which has been one of the 'most ethnographically studied regions in the world' (Carlsen 2011: 47), and that, as such, has developed an ethnographic tradition with many strongly established features that are inevitable to comment upon before trying to advance the idea of cosmology that this work deals with. Three of the topics that this book takes a critical and perhaps unusual distance, given the aforementioned regional ethnographic tradition, are the following: 1) the Wolfian idea of community, 2) the idea of religion, 3) the overwhelming importance of historical contexts and explanations. What I am referring to with 'taking a distance' does not mean an outright denial, but rather a cautious approach to the totalizing character that those ideas have had in regional models, something anticipated in the preceding account of traditional interpretations of cosmology in anthropology.

In general, ethnographies dealing with Guatemalan Indigenous peoples have been 'community' oriented (Brintnall 1979; Carlsen 2011; Cook 2000; La Farge 1947; La Farge and Byers 1931; MacKenzie 2016; Mendelson 1957, 1965; Saler 1960; Tax 1953; Tedlock 1992; Wagley 1949; Watanabe 1992; Wilson 1995). Specifically after Wolf's seminal work (1957), most regional ethnographies have repeatedly discussed the meaning, implications, relevance and even what we could call the 'afterlife' of the term 'community', in the specific Wolfian sense of a 'closed corporate and peasant' community. The continued regional acceptance of the model contrasts with its aforementioned crisis in Mexico, and with the critical re-examinations of its validity in Asia, where authors like Terry Rambo convincingly argued that 'attempting to classify all peasant societies in terms of their position on any single *continuum*, whether it be one that runs from *gemeinschaft* to *gesellschaft*, folk to urban, or closed to open, is not likely to be a fruitful exercise' (Rambo 1977: 187). Therefore, this ethnography will dispense with the concept as an interpretive framework, and merely conserve 'community' as a descriptive label, without any further assumptions.[5] Similarly, while this book will be centred in the K'iche' Maya town of Momostenango, it will not be the study of a community per se, in the sense of a presentation of an abstract collective that holds a number of 'beliefs'. Instead, it will focus on more personal narratives and show their engagements or intersections, as members of society, with the cosmos. Thus, this will imply abandoning the equalization or flattening of cosmological 'data' regarding 'beliefs' that is characteristic of traditional Mesoamerican ethnography, and returning to more individual portraits that show the concrete lives of people and their ideas in their actual contexts, without abandoning, of course, a certain general sense of social validity to them.

A second notion or topic that this book will distance itself from is religion. Religious change and revivals, as well as interreligious conflict, have been crucial topics in ethnographies of Indigenous Guatemalan communities past and present (Brintnall 1979; Hoenes del Pinal 2022; Mendelson 1957; Wilson 1995), especially given the rise of Protestantism in the country during the twentieth century (Stoll 1991). However, while Catholicism and Protestantism (the two most important religions in Guatemala) are indeed religions, the cosmological conceptions and practices that this book will refer to, called *costumbre* by Indigenous people, do not really constitute a 'religion' in the Western sense, since they do not refer to the 'supernatural', another concept that is conspicuously absent in Indigenous cosmologies (Hallowell 1964; Saler 1977, 1993). This important idea has been recently emphasized in the pioneering work of James MacKenzie in the K'iche' town of San Andrés

Xecul, who, following Viveiros de Castro's ideas (1998), preferred to construe *costumbre* as an animistic and shamanistic ontology in his work (2016: 64). Following MacKenzie's appreciation, and since this book is centred in *costumbre*, the religious-centric perspective of past works will be discarded here, and my analysis will not operate from a 'religious logic', nor privilege religion as an explicative tool. In general, I agree with the observations of Daniel Dubuisson (2019), which argue that religion is a concept that was imposed on non-Western people through European colonial projects, and has been ostensibly used as a tool of colonialism, as well as with the opinion of Timothy Fitzgerald, who argues that the category cannot be an epistemological tool, but rather should be an object of critical analysis itself (2000: 106).

Finally, this book will also, in general, forego ethnohistorical chapters and explanations, and mostly bypass the importance accrued to pre-Hispanic symbols in the interpretation of current Maya rituals in certain prominent works,[6] even if, as the reader will see, the sixth chapter of this book is actually devoted to a historical topic. On one hand, the prominence of ethnohistorical overviews in many ethnographies (Carlsen 2011: 71–120; MacKenzie 2016: 34–42; cfr. Tedlock 1992: 13–22) in the region is natural and even desirable, since many Maya communities are still missing detailed ethnohistorical studies. However, in the case of Momostenango, the monumental work of Robert Carmack (1995), one of the most important representatives of the ethnohistorical discipline itself (1972, 1981), makes this endeavour somewhat redundant. But there is something more to this than the issue of the availability of studies and materials, and that is more related, perhaps, to the traditional 'conflict' between history and anthropology (Bayly 2018).

Indeed, recent ethnographies have criticized the overwhelming place accrued to historical precedents in Mesoamerican ethnology, as Alessandro Questa has done in regards to the *masewal* people of Tepetzintla, Puebla (2017, 2023), questioning the way in which the colonial image of Aztec religion, transmitted by Fray Bernardino de Sahagún and continued by authors like Alfredo López Austin and his followers, has overdetermined our vision of Indigenous cosmologies, annulling ethnographic engagements in a desire for standardization and canonization along rigid historical lines (Questa 2017: 182). Indeed, while ethnohistory was originally born out of a desire to give prominence to Indigenous voices and agency in order to defy Western projects of extractivism and dispossession (cfr. MacMillen 2009), in later times the tendency of ethnography in the region has been to understand Indigenous communities as 'subjects' of myriad historical processes beyond their control or, at most, to characterize them as people who 'resist' history, or constantly 'adapt' to it, be it in regards to the Spanish conquest and colonization, the national states or globalization (Watanabe and Fischer 2004). In this regard, this book will emphasize, in Chapter 6, the Indigenous notion of history, which is not deterministic, is cosmologically centred, is active and in which Indigenous agents take the centre stage. Thus, I consider it important for this work to portray this specific idea of history, in order to avoid the inherent deterministic nature of the Western idea of history as applied to non-Western societies, which the ethnohistorical model dominant in regional ethnography has contributed to canonize and perpetuate.

The Idea of Cosmology in This Book

Returning to the main topic of this book, it is possible to ask: What is the vision of cosmology that this book tries to reflect upon? How does it contrast with the aforementioned dominant conceptions in local ethnology? While a complete answer cannot be offered at this point, perhaps an indication could be made through a particularly rich quote by Edmund Leach. Reflecting on the difference between cosmology as conceptualized in Western tradition and Western science vis-à-vis the cosmologies that anthropologists often find in the field, Leach makes a poignant and evocative observation that is worth quoting in full:

> The traditional (religious cosmology) of Western civilization has become frozen by its literary form into a pattern which is wholly anachronistic. A deity which is the 'King of kings, the only ruler of princes' made sense in the days of the Emperor Constantine and even in the England of the sixteenth century, but in a world cluttered with formulae about the equality of man and institutions such as the United Nations Assembly, the imagery of God as a Supreme Emperor has no place at all. Alternatively, the modern (scientific) cosmology of the astrophysicists, with its fables of the beginning of time in a Big Bang and a universe filled with inconceivable numbers of entities spaced at inconceivable distances of time and space, leads us to think that the cosmos is, in all respects, vast, so that the 'other world' if there were such a place, would certainly be a very long way away.
>
> But the cosmologies encountered by anthropologists in the field are quite of a different kind. It was Vico who observed that, for the Greeks, the Underworld was no more distant that the bottom of a plough furrow, while the abode of the gods on Mount Olympus was at the top of a visible mountain of quite moderate size. In the course of my Kachin fieldwork I was told of a procedure whereby a spirit medium, in a state of trance, would ascend a ladder into the sky to consult personally with the sky deities (*mu nat*); the rungs of the ladder were sword blades with the sharp edge upwards. When

> I came to witness the actual performance of this miraculous event I found it something of an anti-climax. The medium ascended on to a platform about twenty feet on the ground and, although the rungs of the ladder did include one upturned sword blade, I noticed that the medium took good care no to step on it! (Leach 1982: 213–14)

In this quote, we can see the main dilemma of cosmology as understood by social anthropology, laid before our eyes. Out account of Western medieval cosmology, which is actually somewhat still the mould of many of our accounts of 'traditional' cosmologies (the aforementioned treatment of Indigenous cosmologies of Mexico and Guatemala being a prominent example) is wholly 'conventional' and 'frozen'; in contrast, the modern image of cosmology conveyed by Western science is less centralized and rigid, but still completely representational. It talks about the totality of the cosmos as a map of a faraway place that remains somewhat alien to us, despite ourselves being actually there. To introduce a third possibility, Leach mentions Giambattista Vico's depiction of actual Greek cosmology, not that of philologists or historians, but that of ancient Greeks *themselves*: for them, Hades was a place that was a bit further down the plough of a farmer, while Mount Olympus was a real mountain to be found within the borders of Thessaly and Macedon. Following this example, but introducing a further complexity that takes us closer to the dilemma, Leach offers a quaint image of Kachin cosmology, presenting almost before our eyes not another instance of a frozen, erudite cosmology, but the cosmology that he found *in the field*: he offers a vision of the Kachin cosmos as it emerges in the theatrical performance of a seer. In this performance, the cosmological ladder that leads to the sky, filled with sword blades placed as rungs, is an actual ladder built on a platform located in the village: the sky itself (*mu*), which is the supreme dimension of the universe in which the gods who are the ancestors of the Kachin clans live, is simply that space that is at the top of this seemingly all-too-mundane platform standing on our earth (*ga*).

That a celestial stairway is somewhat identical to an old ladder is a situation similar to the problem that this book will strive to explain: in summary, the cosmology that we find in the field, not our many models of cosmology. Thus, the questions that this book deals with are the following: 1) Why do people in so many societies often say that their actions proceed not from themselves, but from ancestors, gods and other cosmological entities? 2) Can a non-reductionist understanding of cosmology be proposed for social anthropology? 3) How to engage with cosmology while avoiding the classical understanding of the cosmologies of others as being subordinated to the theoretical activity or the cosmologies of the anthropologists? While my proposal cannot be fully brought forward now, something can be advanced here. Unlike in most of the traditional perspectives reviewed in this chapter, cosmology is not considered here merely as a 'mental product'. It is not a mental representation, a map of the cosmos, nor the 'product of a mentality'. Instead, it is a configuration that emerges when the idea of an 'image of the world' undergoes certain processes that attempt precisely to overcome the difference between totality and representation. It is not a product of society; it is where society takes place. It is not an ideological superstructure in the Marxist sense; it is more akin to a thinking machine. Cosmology is not 'mental', rather, it is 'mind-like' in the sense of Douglas Hofstadter (1979, 2007); that is, a 'strange loop' where the world is twisted back unto itself in order to be able to think about itself.

Summary of the Chapters of This Book

As it often happens in the field, where at first the anthropologist is lonelier and is struggling to find acquaintances and his own place within the community where his work takes place, the nature of the writing of this book will change from an initial ethnography where the voice and preoccupations of the author are more prominent, to an ethnography where other voices become more dominant, to finally approach the more 'objective' or collective topic of historical memory. In the first chapter, I will present some findings and ideas regarding the nature of the sacred altars of the K'iche' Maya, which, as I endeavour to show, are a true nexus for Maya cosmology and whose real meaning was not completely understood in past ethnographies. In Chapter 2, the focus will be on the life of a K'iche' patrilineage, the Ramos family, and their relationship with ancestors and society. Chapter 3 will be dedicated to the celebration of the patron saint of the town, Santiago, but in reality this celebration is framed in a rather complex ritual involving paradoxes related to time. Chapter 4 will be dedicated to different testimonies regarding sorcery and 'witchcraft', which, as I will argue, constitute a sort of 'anticosmology', an alternative idea of causality. Chapter 5 will be dedicated to the portrayal of an Indigenous political activist, Miguel Ángel Pelicó, and his struggle against mining companies and extractivism in the town. Finally, Chapter 6 is dedicated to narrating and understanding the local vision of a historical event: the war of 1903 between Guatemala and El Salvador, in which troops from Momostenango played a vital role, but also in which the cosmological ideas of the K'iche' had an unexpected agency. Finally, the conclusion will reflect on how the cosmological engagements of this book can help us rethink the idea of cosmology in anthropology itself.

Notes

1. The term 'Mesoamerica', originally proposed by the ethnologist Paul Kirchhoff (1943), refers to a historical 'cultural area' comprising part of the territories of the current nations of Mexico, Guatemala, Honduras, Belize, El Salvador, Nicaragua and Costa Rica. The pertinence of the usage of the term to refer to Indigenous peoples of the post-Conquest period has been debated (cfr. López Austin 1990: 32–35), as well as the general utility of the concept itself, since it tends to create a reified, static image that has become problematic in recent years (cfr. Neurath 2023: 9–16); however, I will use this term instead of referring to current national boundaries as identifiers because not only many of the particular aspects of the cosmological conceptions discussed in this book are shared by communities of both Guatemala and Mexico (and within Mexico, the state of Chiapas in particular), but also because the term plays a key part in the debate regarding local Indigenous cosmologies past and present.
2. The most important among them being those of Leonhart Schultze-Jena, who mainly dealt with cosmology, mythology and language in both Chichicastenango and Momostenango (1933); the Guatemalan ethnologist Antonio Gobaud Carrera, who dedicated a brief article to the main Indigenous festival of the town, Wajxaqib' B'atz' (1937); Robert Carmack, who dedicated an authoritative volume to the ethnohistory and the evolution of the sociopolitical organization of the town (1995), as well as an important article to its 'traditional cosmology' (1998); and Carmack's disciples, Barbara Tedlock, who dealt with calendrics and divination (1978, 1992), and Garrett Cook, who dealt with cosmology (1981) and dances (2000). More recent efforts such as that of Thomas Hart (2008) and Paul van der Akker (2018) have continued to increase the ethnographic information available on Momostenango.
3. For a definition of worldview within the Mexican school, see López Austin 2018; for the Mesoamerican worldview as an ideological complex, see López Austin 1988: 16–20.
4. '**κόσμος**, *ὁ*: order ... IV. Philos.: world-order, universe (Liddell and Scott [1901] 1996: 983).
5. Similarly, it is my impression that the idea of a 'death' or 'decline of the community' such as that described by Carlsen in Santiago Atitlan (2011) or by Cancian in Zinacantan (1994) simply masks the reality that the concept, with all the characteristics devised by Wolf, never really existed in its 'ideal' form in the first place.
6. For example, Cook 2000, in which the continuities between contemporary dances and pre-Hispanic imagery is constantly emphasized.

1

'The Law of Us, the Indigenous'

Porque esta es la ley de nosotros, los indígenas:
No de una familia, sino de todas.
For this is the law of us, the Indigenous:
Not of a particular family, but of all of them.

—Marcelino de León

It is always difficult to select a starting point for an ethnography.[1] The usual rhetorical device of describing one's own arrival in a town does not always feel of relevance: rarely does someone stumble with the true subject of their study just 'fresh off the boat' or, in the case of Momostenango,[2] fresh off the famed 'chicken bus' that one must inevitably take to arrive in the town.[3] Initially, I had the intention of doing an ethnography about the relationship between cosmology and power in the town, with an emphasis on calendrics; thus, the difficulty of choosing the best way to approach the topic was (apparently) solved by the fact that, as mentioned in the previous chapter, the social life of the town rests on the shoulders of one of its most well-studied institutions: the *alaxik* or patriclan,[4] and particularly on its leaders, the *chuchqajaw rech alaxik* or patrilineage heads,[5] who are also the main ritual specialists of the town and the keepers of its ritual practices. Thus, at first, I figured out that seeking rapport with one of them would be the best approach, which I did, and most of this initial fieldwork informs the first half of this book. However, soon I would learn that this traditional approach had its limitations, which derived not so much from any lack of kindness or generosity from my acquaintances, but rather from my own bewilderment and lack of understanding before the very complex ensemble of ideas and actions that were displayed before my eyes every day.

The very name Momostenango, a Nahuatl word meaning 'place of the fortress of shrines',[6] is suggestive of the main feature of the town in terms of its ritual tradition: the proliferation of spectacular altars made of potsherds and stone. As many authors have already remarked (Akker 2018: 29–32; Carmack 1998: 342; Cook 2000: 98–99; Tedlock 1992: 77–81; Zamora Corona 2020), most of the rituals in the town are carried out at such altars, both private and public, which are called *awas*, a word that means 'taboo'. Informally, they are also called *porob'al* (burning places) or *quemaderos*, in Spanish, since most of the offerings placed in them are burned, giving them a charred appearance.[7] Such ceremonies involve the cleaning of ritual spaces that are covered with pine needles, and the burning of offerings of native incense, candles, wood and foodstuff, while prayers directed to the ancestors or the earth gods (*dioses mundo*) are recited (Figure 1.1.). So prominent and so present are these shrines in the life of the town, that probably there is no day where one cannot find someone performing a ceremony in the hundreds of altars that dot the landscape of the town.

Figure 1.1. A woman prays at one of the shrines in Pak'lom Hill, the centre of Momosteco cosmology, located within the central area of the town. © Alonso Zamora Corona

As mentioned, my fieldwork started by approaching a patrilineage who practised *costumbre*, the name given to the ensemble of native cosmological ideas and Catholicism practised in Momostenango and other K'iche' towns.[8] Naturally, the rituals that they performed at their lands concentrated on the set of shrines that they inherited from their ancestors. Given their central place at all rituals, at some point during my fieldwork I became convinced that these altars were the key to understanding the cosmology of Momostenango, but I was unsure of what they really were, beyond their very well-known types, functions and calendric properties extensively dealt with in the extant ethnographic literature.[9] Despite the kindness and attentiveness of my informants, with time I realized that, regarding those rituals and structures, I did not really understand what was taking place before my eyes. Former ethnographies touched upon some of the meanings of such structures, and yet, for some reason, I found their accounts unsatisfactory: they stated what the altars were used for (in the case of private altars, protecting crops, family members or travellers; in the case of public altars, a whole array of important social, ritual and political functions), but not what they *were* for the Momostecos. My questions, often based on the information suggested by these works, did not lead to anything substantial or new. The explanations that my informants offered were always honest and kind, but in the end they appeared to be oblique and cryptic, more allusive than explanatory. I also understood that insisting too much would do no good. I temporarily admitted defeat. I just did not know whether the altars had 'something else' to them, or that I did not know how to ask the right questions. Often, when discussing important matters related to cosmological ideas, people in Momostenango would refer to the 'content' or 'essence' of a thing (*el contenido, la esencia*). I felt I was missing those.

This chapter deals with the 'essence' or 'content' of those altars, which, in a way, can be considered as the key feature of the cosmology of Momostenango, not understood merely as a 'world view' in the traditional sense mentioned before, but rather as a much more complex outlook than those previously described. While it does not reflect my first experiences in the town (the events described here took place six months after my arrival), it depicts some of the events that were crucial for me to truly understand what was happening in the town, cosmology-wise. It also introduces my theoretical perspective regarding the surprising and often difficult cosmological explanations I was offered while I was there. It all begins in an unexpected place: with my friendship with a Protestant schoolteacher, Abraham Ajiataz Calel.

Abraham

As mentioned before, my state of bewilderment regarding Momostenango lasted approximately six months, until one day I happened to ask the question to the right person at the right moment, despite this person not participating at all in the 'traditional' ideas of *costumbre*. As mentioned in the quote from Edmund Leach found in the introduction to this book, it is important to realize that cosmology is not an image of the world, but rather a *locus* where people effectively live. The people I usually frequented were, unbeknownst to me, in a 'central' place within the cosmos of the K'iche', despite them not being members of any elite: a respected family, engaged in *costumbre*, involved in the affairs of their community. What I did not know at that time was that, to understand the shape of the K'iche' cosmos, I had to start from the cosmological 'margins', the outside. I talk about a 'cosmological' place or location, because, as we will see, among the K'iche', a person (or a very important part of it) does not only exist in the world before our eyes, but also in another place called *El Santo Mundo*, 'the holy Earth', where their ancestors and the gods live; the name in K'iche' is *Juyub' Taq'aj*, 'the mountain and the valley', a 'coupled kenning' or parallelistic expression that suggests the totality of the cosmos as opposed to the human world.[10]

With time, and independently of my preoccupation with altars, I realized that, if I were to make a more significant ethnography, I had to widen my scope, to frequent not only more people in the community, but also different kinds of people. Luckily, the opportunities came in natural ways. One of the most crucial interlocutors of this ethnography is Abraham Ajiataz Calel, a schoolteacher in his late fifties who works at Instituto Kamawanik, an educational project that was originally funded by the Guatemala Friendship School Foundation (cfr. 2010), providing the funds for the construction of a school that has since fallen into a state of progressive disrepair. I originally sought Abraham in order to further my knowledge of K'iche' and to hire him as an assistant to work with in the translation of Indigenous texts; in time, he became more of a friend and colleague than a teacher, a friendship that endures to this day. We would spend many hours at the ruined classroom where he still teaches many students, talking about his life and the culture and traditions of Momostenango. Born into a *costumbrista* family, Abraham eventually left the traditional religion and converted to evangelical Protestantism. In his youth he pursued an education in pedagogy and psychology, and his skills as a teacher of both Spanish and K'iche' have proven quite useful for him, for he has managed to befriend many of the outsiders that come to Momostenango for humanitarian purposes. Nonetheless, despite the income from his business and the support of two sons who emigrated to the United States, he is usually in financial distress, especially regarding the difficult issue of keeping afloat the school. At that time, he certainly did not partake of *costumbre*, but thanks to his interest in language and his own life experiences, he had a peculiar relationship with the *costumbrista* aspect of Momostenango itself: he felt proud of its traditions and culture, but at the same time broke off with them at a personal and religious level.

As collaborators in the translation and analyses of many hours of recordings, my relationship with Abraham was not that of a researcher with an informant, but somewhat that of two researchers on Momostenango working together. Indeed, Abraham is obviously very knowledgeable about Momostenango's traditions but, as he explained, his parents were not old-time Momostecos, but rather newcomers to the town; Abraham's father was a butcher from Chiquimula who made his living by selling meat at the market. In the context of Momosteco tradition, this means that his family did not have the ancestral altars described above, nor they were members of a traditional patrilineage. His family just kept the altars already found in a property that they acquired, making offerings at them according to *costumbre*. The members of the family would pay a calendrical priest or *chuchqajaw* from another clan to make offerings on their behalf, as some people do. Thus, while Abraham and his family certainly practised *costumbre* before his conversion, he admits he did not know much about it, but he believed in it,[11] and, even more, he told me that, for him, the altars were indeed powerful, for he and many others had witnessed their wondrous powers many times. As he often told me, quoting the old Spanish saying: *no hay que creer ni dejar de creer* ('one should not believe nor cease to believe'), meaning that the power was real, but he did not participate in the ancestral cult anymore.

As mentioned, Momostenango is a 'town of altars'. Indeed, not far from the school where Abraham teaches, a small hill is located, a hill filled with ancestral shrines that was so near us that we could see it from the school's balcony. I once asked Abraham about it, but he admitted he did not know much either. We went there once or twice after working on some recordings, and Abraham proceeded to explain some of his childhood experiences regarding it, for his home is very near to the hill. Being relatively close to the urban centre, the hill is now encroached upon by several buildings. The owner of the place had walled it off. At the top of the hill lies a small cluster of ancestral altars, which some call 'San Antonio Mundo' in honour of a local saint,[12] although some people call it 'the place of the sixty-eight *nawales*' or 'sixty-eight spirits'.

Diviners and calendrical priests would burn offerings there on behalf of their clients; that is why, in a more informal way, these altars are called *porob'al*, 'burning places', or *quemaderos* in Spanish. Abraham did not know much about the place but, when he was a child, his father told him that 'a snake lived there'; not a normal one, but a spirit or *nawal*, and that the animal could be seen in its vicinity sometimes, coiling around the many altars on the ground. When asked about who or what this snake was, he only replied: 'The snake is the owner of the shrine.'

At that point, my enquiries seemed to have hit another dead end, but Abraham, ever so generous, told me that he had a friend, another Protestant, Marcelino de León, a man who had for most of his life been an *ajq'iij* or calendric priest, and that he was quite sure that Marcelino knew much more about these altars and their significance. Marcelino had ceased to burn offerings in his old age, first becoming a Protestant and then leaving active Protestantism because of his personal differences with religious leaders. As is the case with many other family heads at the town, Marcelino usually works on his fields, located in a rural hamlet in the outskirts, but also engages in commercial activities in the town centre, selling many kinds of wares at a street stall placed in the market in front of the main plaza or in its neighbouring streets, depending on the day. Thus, Abraham promised to ask Marcelino about the significance of these altars the next day at the market, and come back to me with a reply.

The explanation offered by Marcelino to Abraham was completely unexpected, and I believe that probably the fact that it was an Indigenous person directly asking another, instead of a non-Indigenous person conducting an interview or doing participant observation, was key to him being more frank and open about it, especially since, as I have explained, no one offered me a similar explanation during my previous months of fieldwork in the town, even if everyone later admitted the information to be actually true.

Marcelino

After being approached by Abraham at his market stall after a long day of buying and selling, when asked about the real significance of the omnipresent altars of the town, Marcelino recalled that many decades ago, when he was just a child and he began being trained as a *k'amal b'e* or ritual specialist by his father, he revealed to him the real reason why people burn offerings at the sacred altars. The reason is that his soul and those of his family are *inside* the altars. 'All families are there, even yours', Marcelino told Abraham. These souls are 'recommended' before the ancestors and the earth divinities, the *dioses mundo*, and thus are protected by them. However, what happens if one does not make offerings at them? Marcelino's father told him that his soul is like a little animal (*chikop*), perhaps a bird or chicken, which lives inside the altar, which represents a corral or animal pen (the word in K'iche' is *korral*, a Spanish loanword), which is placed in the underworld, the *Santo Mundo*, the place where ancestors and earth divinities dwell. When people do not make offerings to their ancestors and the earth divinities, the door of this corral gets opened, and wild animals (*awajmundo* or 'animals of the underworld') enter and start scratching, biting and finally devouring this little bird-soul (Figure 1.2.).

The act of making an offering is seen as some sort of 'closing the door', so the wild aspect of the gods, their *nawal*, will not enter and devour people: 'If a family is not "recommended" in the *awas* … they are like chickens whose enclosure doesn't have a door, there is no protection … The shrines are divided by communities or surnames, for example, all the Ajiataz people are there, and if you were Ajanel, all the Ajanel family members are there', Marcelino told Abraham. This placing of people and their 'names' (*b'iaj*) inside the *awas* is compared to some sort of writing (*tz'iib'am*) or a list (*nómina* in Spanish), for it is said that the ancestors have such lists in the underworld, which state who is to be protected in there, and people who are not protected are 'erased' (*esaj*) from the list; thus, it is said that a sacred altar is truly a 'lord of names' (*rajawal ri b'iaj,* or *el dueño de los nombres* in Spanish). As stated in the opening quote to this chapter, Marcelino characterizes this secret knowledge as *the law* of Momostenango; others would like to say, as mentioned above, that this

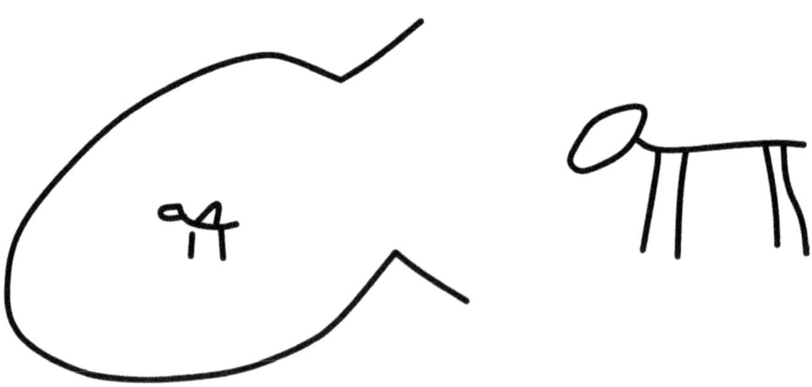

Figure 1.2. A coyote spirit (*awajmundo*) attacking the *chikop* or 'bird soul' of a person inside its enclosure or corral. Drawing by Abraham Ajiataz.

is 'their content, their essence'. This 'content' is not to be disclosed easily to non-Indigenous strangers; that is one of the explanations behind the name *awas*, 'taboo' being given to them.

Marcelino would further explain to my friend that it is not only the *dioses mundo* or earth divinities who menace and devour the souls of people when they are 'outside' the shrines, but even the ancestors themselves, who are said to become 'hungry' when neglected. Thus, if ancestors become hungry, their altars turn into beasts of prey again; they become alive and 'bark and bite' (*kationik*) people:

> For the ancestors have left your name in this altar, and you are not sustaining it, you don't feed them, so they become hungry and fight, that is why you have accidents, failures, illnesses, the snakes arrive, the jaguars, the wild animals, you know why? Because you are recommended to the ancestors and, if you don't sustain them, they bark, they bite, the ancestors bark and bite. If you don't feed these altars, the altars bark and bite. That is why you will have illnesses, you will have failures and death, until you feed them, their food are the offerings, the copal, the candles, that is why you make an offering or pay a *chuchqajaw* to do it ... These altars listen to us and they also speak to us, and they are listening everything we are saying now.

Marcelino finally told Abraham that, despite him not burning offerings any longer, he is convinced that his soul still resides within his altar, for his father 'placed it there' when he was born. This ritual cannot be undone; thus, he attributed the bad happenings of his life to the fact that he did not present offerings at the altar anymore. When Marcelino decided to leave *costumbre* and become a Protestant, he complained to his father thus: 'Why did you put my name there?' And his father replied: 'I did it to protect you.' Thus, 'placing the name' of a person is both an act of protection and an act that makes people vulnerable to the wrath of the earth gods, for it is an act that places your soul in the ancestral world, so to speak. During his conversion, the Protestant missionaries told Marcelino that his ancestors were not earth gods, nor 'lords of the face of the world', but demons. They urged him to throw away the altars, to destroy them. But despite their exhortations, he was impotent to do it. He knew a neighbouring couple that had done so. One day, they took the potsherds and stones out of their property and destroyed them, smashing them into pieces. The vengeance of the ancestors was swift. The couple used to travel to the coast to do business. One day, both fell into the river, and they drowned. The people of the vicinity agreed: their deaths were the work of the ancestors. This is what is known as *kekationik ri awas, kekationik ri nantat*, 'the altar, the ancestors bark and bite'.[13]

Later, I arranged to personally meet Marcelino and he confirmed to me the details of his narrative, while showing Abraham and me the lands of his patrilineage and the different altars still there, all abandoned. During that occasion, Marcelino explicitly told me that his spirit was inside one of such shrines, called *warab'al ja* ('house of dreaming') by pointing at it and telling me, 'I am placed here' (*yo estoy aquí metido*). He told me that when you become member of a patrilineage (either by birth or by marriage) your spirit or essence must be placed inside by being presented to the ancestors,[14] and when the shrine becomes abandoned, it can kill you:

> The altar, if not taken care of, it bites, if you cannot sustain it, it can kill one, it can kill quick. If you don't feed it, it kills you, if there are no candles, it kills. Because it wants its spiritual nourishing. My father put my name in the *winel*, because he was a follower of the ancients, he put my name there ...

For example, when a woman marries a man, her name must be placed inside the ancestral altar of the man's family.[15] If she, for example, tried to escape or bring dishonour, 'she would only find death, for she cannot ever leave. She is here from now on. She will be found, whenever she is, and will be undone'. He used myself as an example: 'If I place you here, you are not going back to Mexico anymore. Not anymore. You stay here.' After this, Marcelino told me that, during the nights, the *nawal* of the altar rises like an animal, goes out, bringing illness and death, for it hungers for offerings:

> Marcelino: And if I don't give it sustenance, food, then the neighbours will die, it will sting another, any failure will happen, because there isn't food there anymore, it bites you ...
>
> Abraham: It bites, they (the *costumbristas*) say 'it bites'. Now, if you don't give it sustain, food, it bites you ...
>
> Marcelino: If one doesn't 'work' it, doesn't offer candles ... The altar will devour you.

Thus, feeding the altars is the duty of the calendrical priest, on behalf of his family and of the community. He reads the omens that the earth gods send to people either through divination or through the 'lightning' in his blood,[16] and perform the adequate ceremonies. When ceremonies are performed correctly, the people are 'wrapped' (*keb'otz'ik*) inside the altar,[17] as if they were embraced by it, being tight and protected within. But Marcelino would not do ceremonies anymore.

One day, Marcelino took us on a visit to the great mountain shrine called Pipil Ab'aj (Figure 1.3.).[18] He offered us many interesting explanations

Figure 1.3. Marcelino de León at the mountain shrine of Pipil Ab'aj. © Alonso Zamora Corona

regarding the functions of the place back when he performed rituals on it; however, he gave us one of the most interesting testimonies about it when we were in our way for the pickup to take us back home. During our walk, Marcelino started talking about the *awajmundo*, the 'animals of the earth gods'. Once, his father saw them, in a small cave near a shrine called Campana Mundo, 'the bell of the underworld', named so because the sound of a bell is said to be heard nearby at night.[19] His father was on that occasion with a friend, and both entered the cave. Once inside, they found a pair of coyotes, who proceeded to speak to them and told them that they had incurred a great offence by seeing them, for they were none other than earth gods themselves. Later, the friends duly paid for this offence with offerings to the lords of the shrines that they had seen in their animal form, an act that was also taboo in itself:

> … Our ancestors tell how that is a place of great significance, too. For a pair of coyotes were born out of the stones, my father saw them, when he arrived one evening, oh god, he was reprimanded by them, they took a 'fine' from him, and he who guided him had to pay, too. 'Why do you show him these things?' they said to him, and then they paid to them a 'fine' (*multa*), their present, they were saved, yes, but the animals didn't like it, the animals of the god of the Earth (*awajmundo*) …

During the conversation, Marcelino explained that ritual precautions (also called *awas*) exist precisely because of the wild, vengeful character of the ancestors and the 'lords'. For example, sexual abstinence before offering at a shrine is obligatory because, if a calendric priest were to touch his wife before coming to an altar, he would be somewhat placing his own flesh inside his offerings to be consumed: 'It is as if I place myself inside the incense' (*vengo a ponerme yo mismo en el copal*) he told me. Therefore, he would be slowly devoured by the gods, so he would become ill, have bad luck, fall in a river and so on. The prohibition only extends to the day before doing 'the working'; after that, you can do 'whatever you want', as he told Abraham and me while laughing at the innuendo.

As an example of the consequences of breaking a ritual taboo, Marcelino told me about his father's undoing. He recalled that his father was a calendrical priest, but at one point he 'failed'. He told me that, when doing workings at an altar, you must 'open and close them', on specific days of the Maya calendar. Thus, first one must 'open the door' to communicate with the earth gods and ancestors and ask for favours. During that time, one, and the people on whose behalf one makes offerings, must take some ritual precautions, including sexual abstinence. If not, you will be seen by the earth gods as food. Then, after the blessings are received, the altar is to be 'closed' through another ritual (he would not disclose the precise days, which strangers are to ignore). The problem is that his father 'forgot' to perform the ceremony of closure. The *awajmundo* spirits began attacking people, since people did not 'take care of themselves'; that is, they fell into ritual impurity. Bad luck ensued, and people regarded his father as an untrusty calendric priest. That was his social undoing, which prompted Marcelino's conversion.

A final comment regarding Marcelino's conception of his own conversion is needed. As Marcelino told me and Abraham when we visited his land, being 'inside' the altar of the *Santo Mundo* is something that cannot be undone. Thus, he had to deal with the *nawales* or spirits for the rest of his life. He would pray, fast and perform other spiritual Protestant practices to resist their power. But, despite this grim outlook, he had a secret hope that he confided to us on the day he showed us the altars of his lands. His grandson, who was looking at us while we were walking through the *milpa*, was never 'placed' inside the altar, for he was born after his conversion, after he ceased doing offerings. Thus, he was to be 'free' of the 'law'. His conviction about the freedom of his child kept him going, despite his personal disagreements with the leaders of the church, whom both Abraham and Marcelino characterized as excessively competitive and domineering at times.

Having presented the fascinating story of Marcelino's conversion and his deep account of the Indigenous cosmology of Momostenango, it is time to discuss its implications for the idea on cosmology that this book proposes.

Looking Back into the Ethnography of Momostenango

Momostenango has been the subject of intense ethnographic research during the past century, mostly motivated by its reputation of 'traditionalism'. Perhaps the most important work dealing with the institutions of the town, past and present, is that of the ethnohistorian and anthropologist Robert Carmack. The main products of his fieldwork in Momostenango are a series of seminal articles on the town's history (1979), a thorough ethnohistorical monograph (1995) and an article that outlines the 'traditional' aspects of social life in Momostenango (1998).[20] In general, Carmack defined his approach as 'historical anthropology', for it was both informed by fieldwork and archive research (1998: 324).

The work of Carmack was continued, but also expanded and partly revaluated, by his students, Garrett Cook and Barbara Tedlock. Cook prepared an important and, sadly, unpublished comparative account of K'iche' cosmology as his PhD thesis (1981) and later wrote what can be considered as the definitive monograph on *cofradía* or sodality cults in Momostenango from a symbolic perspective, highlighting continuities with Classic Maya conceptions (2000), considering current K'iche' rituals of the present as means to reaffirm and reassert complexes of ancestral symbols. Tedlock, who is perhaps more famous than her master nowadays, resided in the town between 1975 and 1976, and had subsequent sporadic stays before the worst period of the civil war started in 1982, which forced her to leave town, only being able to return many years later. Her main concern was the study of calendrical divination in the town, which was the subject of her PhD thesis (1978), and became the basis of her classical monograph on the subject (1992). Being initiated as a calendrical diviner herself by the Xiloj patrilineage, Tedlock's account is the basis of all posterior treatments on the subject. However, Tedlock's treatment of the topic created a sort of theoretical rift between master and disciple, since Tedlock's holistic ideas on K'iche' divination radically departed from Carmack's more sociological, Weberian perspective.[21]

In regards to the information presented in this chapter, the 'corral' cosmology described by Marcelino is certainly new regarding our current knowledge of Momostenango, since neither of the ethnographers mentioned were informed in detail about it.[22] However, similar cosmologies are well known in the context of other Highland Maya groups, especially in Chiapas, where conceptions akin to the ones mentioned in this chapter were first described in different works by William Holland (1961, 1964), Evon Vogt (1969) and Gary Gossen (1975), now considered as classics in the anthropology of Maya groups. Such works inspired and shaped the classic anthropological model discussed in my introduction; that is, the structural-functionalist school in Mexico. Certainly, some core elements described by the classic anthropology of Chiapas are present in Marcelino's account too, like the relationship between humans and animal spirits, which determines the social standing of the individual; that is, the relationship between human society and the 'society of the sacred mountain', as Holland termed it (1961: 170). In fact, it would be easy to go back to such structural-functionalist explanations, based on an isomorphism or homology between cosmology and society. Indeed, with these materials one could construe a 'structural-functionalist Momosteco cosmology', considering these ideas as mere reflections of social functions.

However, as also discussed earlier, it was precisely the criticism of these ideas that gave rise to the modern debate on Indigenous cosmologies in Mesoamerica, and their explicit rejection and criticism has become something of a *raison d'être* for ethnographies of following generations. Such authors, like Pedro Pitarch (2010) and Helios Figuerola (2010), decided to concentrate their efforts instead on native notions of the soul, the world, and the human body in themselves, rather than conceiving them as shadows of a social order and its 'structures', producing detailed, engaged accounts of such concepts. It would also be tempting, in the context of this ethnography, to produce a similarly oriented work, discussing in detail such ideas in and of themselves, and comparing them to similar systems in the wider context of the Chiapas-Guatemala region, a work that certainly needs to be done from a strictly ethnological point of view.[23] Tedlock's (implicit) criticism of Carmack's description of Indigenous practices in purely Weberian notions of social power, and her personal and anthropological search for the ways in which Indigenous people overcame Western dichotomies could be seen as anticipating this tendency, too.

However, the present ethnography also distances from the recent tendency towards discarding the social dimension of cosmological conceptions that characterized more recent ethnology, although it recognizes it as a necessary strategy in the ebb and flow of anthropological thought. This ethnography is based precisely on the idea that a return to the social is necessary, although this return needs to go deeper that old functionalist views did, and take Indigenous ideas seriously. As Martin Holbraad has suggested regarding the idea of a 'post-critical anthropology', the ontological turn has dared not only 'to mention, but actually use' in a theoretically novel way the ideas and concepts of several authors of the anthropological past who were dismissed in a categorical, 'scorched-earth' kind of way during the numerous crises of the discipline (2017: 285). As discussed earlier, such crises took place also in Mesoamerican ethnology, with lasting consequences. In general, the classic obsession with social 'structures' and 'patterns' was eschewed, along with attempts to link cosmology and society. However, as it was explained to me time after time, the main reason why people in Momostenango engage with cosmological concepts is to either 'fix' the course of their lives or to ensure the success in business, family life or political and social endeavours.

Thus, I would suggest that this entanglement of the social (the concrete) and the cosmological is not only a prominent phenomenon in Momostenango, but rather *the defining element of a true cosmology*, which otherwise would indeed be merely an abstract 'world image' or a 'world view', instead of a real place filled with living beings of all kinds that have concrete effects upon people's lives and upon society. Furthermore, something akin to this entanglement was somewhat prefigured in the work of classic ethnographers of the Maya, despite the now obvious criticisms we can leverage at their structural-functionalist theories. What is remarkable, though, is that this intermeshing is clearly defined by the K'iche' themselves in their altar cosmology, which, more than mere ritual structures, can be seen as being somewhat of a theory of cosmological entanglement, whose description is the object of the paragraphs that follow.

A Cosmological Entanglement

As the general orientation of this book entails, that is that ethnography is the ideal arena for theoretical experimentation, let us forget about souls and 'worldviews' as ethnological objects and return to the scene described before: that of Marcelino explaining to me how he is inside his altar, and try to formulate its complexity in more theoretical terms. When Marcelino says: *Yo estoy aquí metido* (I'm inside here), pointing at his altar, he is asserting a paradox, for I can see him standing before me, and yet he says that his 'soul', which is 'inside him', is also contained within another thing that is external to him, the shrine. This shrine is also the outer form, the container of the dwelling of his ancestors in the *Santo Mundo*, where Marcelino is ultimately contained too. However, the paradox is that *he can even manipulate this container of himself while being outside of it at the same time*. The proof of this is that, as Marcelino acknowledges, if he destroyed the shrine, *he would eventually destroy himself*, just like the couple that fell into the river; exposed to the wilderness of the desolate cosmos, his *chikop* spirit would be devoured. Thus, cosmologically, he is standing both inside and outside the shrine. This image delineates a loop of sorts, for Marcelino is both inside and outside the very cosmological vessel he is portraying before me; he is contained in it, and yet he is outside of it, pointing at it. The resulting image is that of a paradox of double self-inclusion: the altar contains Marcelino and the world around him,[24] but at the same time, the world where both Marcelino and I are standing contains the altar itself.

A first approach to this ontological arrangement could be suggested by the so-called Droste effect, 'the double-mirroring effect created by placing an image within an image and so on, repeating infinitely' (Chandler 2011: 278), a visual paradox named after an advertisement for cocoa from the Droste chocolate company in the Netherlands (Figure 1.4.). However, there is a crucial difference with my ethnographic account, because the Droste effect is usually arranged in a hierarchically descending manner, a sort of stairway, with the higher level containing an iteration of itself in the lower one, and so on, without an end.[25] In contrast, in the cosmology described by Marcelino, both levels are contained within each other, both occupying the 'first place' of the hierarchy in a looping or cyclical manner.

Figure 1.4. The Droste effect, a picture that contains itself in a recursive manner, and a breakdown of the first three levels present within the picture. Wikimedia Commons, public domain.

The idea of a paradoxical arrangement or interconnection of different levels that contain each other has been characterized as being a 'tangled hierarchy', a notion formulated in the work of Douglas Hofstadter, *Gödel, Escher, Bach: An Eternal Golden Braid* ([1979] 1999) and expanded by the author in later works (2007). According to Hofstadter's basic definition, a tangled hierarchy is 'a system in which a strange loop appears' ([1979] 1999: 10). But what is a 'strange loop'? 'The strange loop phenomenon occurs whenever, by moving upwards (or downwards) through the levels of some hierarchical system, we unexpectedly find ourselves right back where we started' ([1979] 1999: 10). This idea is somewhat difficult to explain in merely abstract terms, because it breaks with everyday logic. Hofstadter's classic work exposes the occurrences and the functioning of 'strange loops' in all kinds of seemingly unrelated phenomena, from Bach's canons, to M. C. Escher's prints, to Gödel's theorems, to biological and computer viruses.

A visual example that contains both ideas that characterize the cosmology outlined above; that is, a tangled hierarchy and a self-inclusion paradox, can be found in Hofstadter's analysis of the lithograph 'Print Gallery' (*Prententententoonstelling*), by M. C. Escher (not reproduced here). In Hofstadter's words:

> What we see is a picture gallery where a young man is standing, looking at a picture of a ship in the harbour of a small town, perhaps a Maltese town, to guess from the architecture, with its little turrets, occasional cupolas, and flat stone roofs, upon one of which sits a boy, relaxing in the heat, while two floors below him a woman – perhaps his mother – gazes out of the window from the apartment which sits directly above a picture gallery where a young man is standing, looking at a picture of a ship in the harbour of a small town, perhaps a Maltese town … What? We are back on the same level as we began, though all logic dictates we cannot be. (Hofstadter [1979] 1999: 715)

In a way, we could take Escher's image as a visual counterpart of the statement *yo estoy aquí metido* ('I am inside here'). The visual paradox achieved by Escher's lithograph is determined by the different kinds of ways that the picture expresses the idea of 'inside': 'The gallery is physically in the town ("inclusion"), the town is artistically in the picture ("depiction"), the picture is mentally in the person ("representation")' (Hofstadter [1979] 1999: 715). The paradox is that we can see the young man 'inside' of himself, by jumping from one level to another. The loop can be collapsed in only two steps, as Hofstadter argues (Hofstadter [1979] 1999: 715). Finally, as Hofstadter asserts, crucial to this effect is the central 'blemish' or blank space where the signature of the artist stands, for without it the self-inclusive mechanism would be impossible. This 'incompleteness', though, is never perceived by the young man, trapped inside the world of the picture; it arises to the mind of the other observer, who lies outside the picture. However, as demonstrated by B. de Smit and H. W. Lenstra Jr in their mathematical analysis of the picture (2003), when filling the white spot with the help of a computer algorithm, they found that the drawing repeats itself indefinitely in an ever-diminishing scale.[26] Thus, the 'white space' must be taken as a device to collapse the ever-descending hierarchy of self-contained images and return them, through a loop, to the first level. Thus, Escher's original picture collapses what is an underlying and potentially infinite structure, closing the loop present within the Droste effect by connecting the first level to the last, combining a self-inclusion paradox with a tangled hierarchy.

Returning to our example, however, there is a single crucial difference, for the images discussed rely, of course, on representational means.

Hofstadter's analysis is ultimately representational, for any picture is so.[27] However, in our case, inclusion is not 'physical' nor representational only, but ontological, what we could call a 'strange ontological loop'.[28] If we compare the man looking at the picture in the lithograph and Marcelino looking at his altar, while the man in the gallery is only 'depicted' in the image that he sees, in Marcelino's point of view *he is truly there*, not merely 'represented there'. Thus, the cycle of inclusion is closed in a loop, going 'downwards' from the cosmos to the altar, and then, once there, moving 'upwards', from the altar to the cosmos (Figure 1.5.).[29] This is the reason why changing something in one level implies changing its self-contained reality in another; what is external is somehow presented as being internal,[30] and, ultimately, the social (and the natural too) is always presented as being cosmological, and the cosmological as eminently social in the numerous testimonies that will appear in this ethnography.

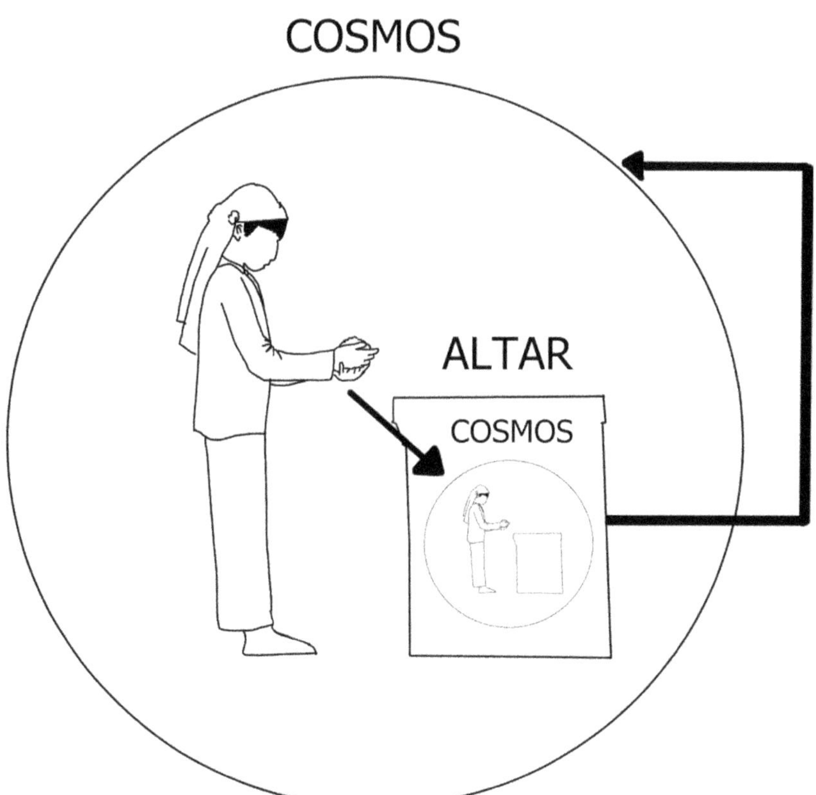

Figure 1.5. Self-inclusion and hierarchical entanglement in the relationship between cosmology and altars in Momostenango. © Alonso Zamora Corona

Thus, a 'cosmology' is something very different depending on whether a Western outsider or an Indigenous K'iche' is conceiving it. For a Westerner, a cosmology is nothing more than a map, a representation, in where people, nature and things are 'depicted'. There is a *correspondence* between representation and reality, but there is no true *equivalence* between them. Or, quoting a phrase that Gregory Bateson was fond of, 'the map is not the territory' (Bateson and Bateson 1987: 21).[31] However, the opposite is true for the Indigenous person. In their point of view, what is for us a map (an altar as a cosmological container), is also a 'territory'. In this territory, their existence, as well as the one of everything in the world, is duplicated. It is as if, in Escher's 'Print Gallery', our conception of cosmology is that suggested by the first half of the scene, where the spectator sees a representation of the harbour where the gallery is placed, and where they are, but without the 'loop' suggested by the second half of the picture, whereas the Indigenous person sees the whole picture from the Escherian spectator, who sees how the image overflows and loops into itself. For this spectator, if they were to tear the image down, they would tear themselves to shreds. For us, this is akin to being 'trapped' into a representation, although this is not precisely the case, since, again, Indigenous cosmologies are not representations.

Thus, the dilemma of Marcelino's conversion is beyond that of changing a 'worldview' for another, changing one representation for another. Because his cosmology is not a 'worldview', since (unlike, let us say, Escher's lithographs) *it is not an image or a representation*, it is a real place, from which it is impossible to escape from, once placed there. And this realization leads us to one of the most important ideas that this book tries to advance: that this 'strangely-looping' structure, this 'tangledly hierarchical' cosmology is the key to understanding how cosmologies in anthropology operate, or at least some cosmologies as opposed to our own, in which such loops are (perhaps) non-extant.

For example, let us recall once more Leach's ethnographic anecdote about a Kachin diviner consulting the sky deities by going up the rungs of a ladder placed in a platform during a ritual performance (Leach 1982: 213–14). If we somewhat project the idea explained in this chapter to Leach's example, we can make further sense out of it. Indeed, if we think the relationship between the sky (*mu*), the earth (*ga*), and the ritual ladder, which echoes a cosmological ladder, we find that, ontologically, this relationship is also a similar loop. The ladder *is* both the ladder at the sky and a ladder on the earth. A seer does climb up the sky by going up it, and talks with the *real gods*, who are sitting at its end. This entanglement, which is performative or 'invented' in the sense of Roy Wagner (1975), rather than only 'given',[32] this double identity of two beings who somewhat contain the reality of each other in a far-reaching 'strange loop' that makes possible what seems to be, following the old physics moniker, 'spooky actions at a distance',[33] is, as I will later propose, the general *form* that makes cosmology work, the ontological structure that defines cosmology as lived, as opposed to any characteristics or ontological determinations contained within a given cosmology, or to ontology conceived as a 'replacement' for the apparently old-fashioned notion of cosmology.

Furthermore, this structure is what differentiates Western religion, more of a 'moral' concept dealing with the 'supernatural', from Indigenous cosmologies, which are ontologically real places from which sometimes it is difficult to 'escape' and have different cosmologies interconnected with each other, being 'multinatural' (Viveiros 1998). The proof is Marcelino's peculiar outlook, considering himself, if we could say so, a Protestant living in a K'iche' cosmology, and his son as a Protestant living outside this cosmology. Marcelino is, using the Aztec colonial term, *nepantla* ('in-between' or 'in the middle'),[34] living in a Maya cosmology with a Protestant outlook.

Finally, as I hope to explain later in this book, this entanglement creates the possibility of what we anthropologists call 'magic', which Nahua people called *nahuallotl* ('hiddenness') and the ancient K'iche', using a term inherited from them, called *pus nawal* ('magic power'), which is, indeed, not a mere operation among others, but rather any action that operates across the whole topology of a given universe, 'non-locally', implying with this more than mere physical distance. For indeed, we will see later, that which Marcelino evoked was not only a condition for the protection or the lack of it of beings within the cosmos, but also the possibility for any true cosmological action.

Notes

1. Instead of preceding my ethnography with a chapter on the specific context of Momostenango and K'iche' ethnology, I have opted to explain their key concepts as they emerge in my ethnographic account. Readers are encouraged to look at these end notes for more detailed explanations and references. Likewise, Appendix A of this book is a summary on basic K'iche' cosmological concepts, and Appendix B a glossary of K'iche' terms.
2. Momostenango, head of the municipality of the same name, is located in the Totonicapán department in the western highlands of Guatemala. According to the last census of the National Institute of Statistics of Guatemala, the municipality has a total population of 105,617, of which 97.7 per cent self-identify as Indigenous and speak the K'iche' language of the Eastern Maya language family; 85 per cent of the population of the municipality live in rural areas (Centro Latinoamericano y Caribeño de Demografía 2019).

3. The colourful decorated buses that are the most used vehicle of public transportation in rural Guatemala, usually adapted from retired American school buses.
4. The *alaxik* or Momosteco patrilineages are corporate descent groups that reckon descent in the agnatic line and keep records of their genealogies and their branches (*xeteil*) for several generations (Carmack 1995: 31). Each patriclan is named on behalf of its founding ancestor: Vicente, Ixcoy, Ajanel, Poroj, Pelico are some examples. The members of a patrilineage, both living and deceased, are called *komon*, a term that usually is used to refer to large gatherings of people and to communities (Pontious 1980: 18). Exogamy among patrilineages is strict.
5. The authority of the affairs of the patrilineage is usually placed on a head or 'chief' called *chuchqajaw rech alaxik*, who has been recognized as having 'a sacred ... association with the ancestors and the earth' (Carmack 1998: 35). He must become initiated as a calendric specialist (*ajq'iij* or 'he of the day'), and as such it is his duty to makes offerings, prayers at special shrines associated with his clan and to perform divination. Due to his expertise as both a ritual specialist and a leader for his clan, a *chuchqajaw* can exert ritual functions on behalf of others and be regarded as a general counsellor for the affairs of his community, becoming a sort of moral authority, a role that is reflected in his other name, *k'amal b'e* or 'guide of the path'.
6. According to the prevailing theory, the current name of the town was given to it by the Conquistador allies, the Tlaxcaltec, during the sixteenth century. In Nahuatl, *momoztli* means a shrine or altar, *tenamitl* means 'wall' or 'fortress', and the suffix -co denotes a place. However, the Nahuatl place name could also be pre-Hispanic, given the frequent contacts between the Aztec empire and the K'iche' (cfr. Robles Castellanos 2007: 215–24), the hypothesis that I favour. The original K'iche' name of the town is Chuwa Tz'aq ('before the monument/shrine'); the Nahuatl name could be an approximation of this expression.
7. While the *awas* are sacred spaces that are better conveyed by the English word 'shrine', most of the time people would refer to them, in Spanish, as *altares*, 'altars'. Thus, both terms will be used indistinctively in this ethnography.
8. *Costumbre* has been characterized as an ensemble of 'ancient principles' (Carmack 1998: 334), as a 'syncretized Maya-Catholic religion' (Cook 2000: 267), and, more recently, as 'an ontology and set of practices that are largely animistic and shamanic in character' (MacKenzie 2016: 64). It comprises both Catholic and Indigenous ideas and practices, such as the calendrical rituals that will feature prominently in this book. As mentioned in the introduction, I will not treat *costumbre* as a religion, but *costumbristas* describe themselves as being Catholic, religion-wise.
9. Besides the public shrines, most *costumbrista* families have a set of private shrine patrilineages in their lands, at least three of them: one called *winel*, for agriculture; one called *warab'al ja*, where the spirits of the patrilineage are enshrined; and one altar for material abundance, called *meb'il*, which sometimes takes the form of a stone with the shape of an animal. Agricultural ceremonies, like the blessing of seeds, are carried in the *winel*, while the *warab'al ja* is reserved for ceremonies regarding the patrilineage, such as the presentation of new children or wives to the ancestors, for the altar is the 'sleeping place' of the souls of the ancestors (cfr. Carmack 1998: 342; Cook 2000: 98–99; Tedlock 1992: 77–81; Zamora Corona 2020).
10. For couplet structures and parallelisms in Mesoamerican literature, see Bright 1990 and Knowlton 2002. The term 'coupled kenning' is a translation of the Spanish *difrasismo* (cfr. Bright 1990: 440).
11. While representational statements like those implied by the word 'belief' are certainly disputed now in their usefulness or pertinence, especially by the ontological

turn (cfr. Holbraad and Pedersen 2017: 17), I mention it here not as analytical category, but as that which Abraham stated. As will become apparent later, this 'belief' is so strong that even non-*costumbristas* are convinced of the power of the ancestral altars.

12. The cult of San Antonio Mundo will feature prominently in the following chapter. For now, it will suffice to say that he is a native version of San Antonio de Padua, patron saint of animals in European Catholicism; however, in Momostenango (where he is usually depicted as a child rather than as a grown-up man), he is the patron saint of the Santo Mundo or cosmological dimension itself, and the wellbeing of families (cfr. Cook 2000: 64–75).
13. Garrett Cook already reported how the destruction of ancestral altars led to 'death or ruination' for families (1981: 11), although without specifying the reason for it.
14. See Estrada Peña 2010 for a description of a similar ceremony.
15. This happens, of course, because the K'iche' are patrilineal and patrilocal, and a woman will always go to live with the family of her husband, and not the opposite.
16. 'Lightning' is the name given to a set of corporal signs felt as a 'tingling' within the body of a diviner, said to be in her/his blood, which is crucial to understand the messages of earth divinities, among which the divinized days of the calendar are included. Tedlock (1992: 138–47) has dealt extensively with the topic.
17. The word denotes being enveloped in a bundle; sacred ritual bundles (often containing the relics of gods and ancestors) have a central place in Mesoamerican cosmologies (Guernsey and Riley 2006), and feature prominently in the ritual life of the K'iche' since Pre-Hispanic times (Christenson 2003). So important are they that Pedro Pitarch has recently proposed a 'textile topology' inspired in the Deleuzian concept of the fold (1993), based in his fieldwork among the Tzeltal (Pitarch 2021); similarly, Daniel Grecco Pacheco has recently studied the ontology behind the concept of wrapping among both Classic and modern Maya (2021).
18. In Momosteco cosmology, it is the sacred mountain of the north (Tedlock 1992: 112).
19. I have been unable to find another reference to this altar or its location. It could probably be Campana Ab'aj, located in the mountains to the west of Totonicapán.
20. Carmack's fieldwork began in 1966 and he continued to be engaged with the community through several later trips, his work being focused on the political and legal life of the community (cfr. 1995: xv). Interestingly, Carmack's memory is alive in the town, especially with some of the older Ladino families in the town centre, because it seems that he lived among them.
21. Tedlock characterized K'iche' divination as overcoming Victor Turner's famous division between 'revelation' and 'divination' (1992: 137; cfr. Turner 1975), as well as Marxist and Hegelian schemes (Tedlock 1992: 196, 206), to propose the idea of a 'dualistic holism' (1992: 201). Later, she developed the idea of K'iche' divinatory rituals as a form of 'shamanic healing' (2005a: 15) and, as a result of her initiation, went on to promote these rituals at different workshops in the United States (cfr. 2005b), describing them as being based on a 'holistic life force' (cfr. Donovan 2005).
22. Interestingly, though, in an article called 'On a Mountain Road in the Dark: Encounters with the Quiché Maya Culture Hero', Barbara Tedlock (1986) mentions a shrine to the K'oxol or Tzitzimit, the god of game and patron of divination, located in another K'iche' community (Santa Cruz del Quiché), where the aforementioned divinity keeps the souls of game animals in a miniature corral and protects them. However, this was never related by Tedlock's informants to the spirits of humans nor to the sacred altars in the community of Momostenango itself. Nonetheless, this

information is of great significance, because, as it will be evident from these and other accounts, it seems that the patrilineage altars in Momostenango somehow mirror the sacred altar in Santa Cruz in many respects and, furthermore, according to another myth I obtained, the sacred altars of the Momostecos are nothing but animals frozen in stone and potsherds, which mirrors the narrative found in the *Popol Wuj*, where the K'oxol escapes the fate of the predator animals that roamed the earth before the rising of the sun (which turned into stone) by hiding in a tree (Christenson 2003: 229–30), another association brilliantly noticed by Tedlock (1992: 148).

23. I have produced an account in such style regarding soul-conceptions in Momostenango and their counterparts in other regions (Zamora Corona 2019); a summary of the Momosteco information can be found in Appendix A.
24. In both the changed animal form of animal spirits inside corrals (*nawal* and *chikop*), and a human form, the *k'ux* or anima, a 'heart-soul' of sorts that, however, is not a mere part of the person, but a copy of the person itself.
25. For an overview of recursion in anthropology and in the sciences of complexity, see Reynoso 2006: 343–6.
26. The website 'Escher and the Droste Effect' contains both images and videos of this analysis (de Smit 2003). Also, see Leys 2007 for a computational application of Lentra and Smith's work.
27. Indeed, in a passage discussing a similar visual paradox (the famous Escher's lithograph, *Drawing Hands*), and its consequences in his theory of the human mind, Hofstadter laments its imaginary quality: 'The abstract structure in *Drawing Hands* would constitute a perfect example of a genuine strange loop, were it not for that one little defect – what we think we see is not genuine; it's fake!' (Hofstadter 2007: 103). But my ethnographic example does not have that 'defect'.
28. 'What I mean by "strange loop" is … not a physical circuit but an abstract loop in which, in the series of stages that constitute the cycling-around, there is a shift from one level of abstraction (or structure) to another, which feels like an upwards movement in a hierarchy, and yet somehow the successive "upward" shifts turn out to give rise to a closed cycle. That is, despite one's sense of departing ever further from one's origin, one winds up, to one's shock, exactly where one had started out. In short, a strange loop is a paradoxical level-crossing feedback loop' (Hofstadter 2007: 101–2).
29. For example, Abraham partly *believes* in the power of the *costumbre* cosmology (a mostly representational statement), but he was never *inside* of it (*inclusion*), or at least he never knew he was. Marcelino is, however, *inside*, but this inclusion is not physical nor representational (it is not a belief), but *ontological*.
30. This idea could be compared to Stefan Feuchtgang's account of Chinese cosmology (1992) and its practices of centring (2014).
31. The phrase 'a map is not the territory' was coined by the mathematician Alfred Korzybski in a text dealing with the complexities and paradoxes that arise from the problem of infinity (2000 [1933]: 750). Discussing the relationship between maps and languages as representations of things, Korzybski observes that a map always has a dissimilar structure to that of the territory it represents even if it has a similar logic structure, since 'an ideal map would contain the map of the map, the map of the map of the map, etc., endlessly …'. He calls this characteristic 'self-reflexiveness' (2000 [1933]: 751). Gregory Bateson, most known for his ground breaking ethnography *Naven* (1958), was fond of Korzybski arguments, and often reflected upon the way in which people 'confused' maps and territories (1979: 30–31); later, the difference

between both would become a foundational element for his theory on the relationship between 'mind' and 'nature' (1979).

32. Indeed, as we have seen and we will see later, the Maya entanglement is performed: as Marcelino said, it happens when the calendric ritualist 'places you' at the calendric shrine.
33. The derogative name that Einstein gave to the phenomenon known today as 'quantum entanglement'; that is, the capacity of quantum objects to affect each other over very large distances (Ball 2018: 11). However, nowadays it is not considered a real action but rather an entanglement of the properties of quantum particles, and is better explained through the concept of 'non-locality' (Ball 2018: 184).
34. In his *History of the Indies of New Spain*, the sixteenth-century Dominic friar Diego Durán explains that the Aztec of his time, recently converted to Catholicism, retained the usage of many pre-Hispanic cultural, ritual and cosmological elements, such as the use of the sacred calendar or *tonalpohualli* (the Aztec equivalent of the K'iche' *chol q'iij*). When asked about it by the friar, an Aztec man replied to him, 'Father, do not be astonished, we are still *nepantla*' (Durán 1971: 410), meaning 'in the middle' of Catholicism and the old ways. The term has been used to describe mixed identities or customs, and is specially favoured by the Chicano community (Scott and Tuana 2017), but it does not necessarily imply 'biological' *mestizaje* nor religious 'syncretism'. In this case, though, the *nepantla* condition would refer to something deeper than 'symbols', and would refer to being somewhat unable to escape the Indigenous cosmos itself.

2

Seeing with Eyes of Fire

A Ritual Seance Among a K'iche' Family

The last chapter explained the deeper meanings of K'iche' altars and the paradoxical nature of the cosmology of which they are a crucial nexus. However, the perspective presented of them was crucially and somewhat strategically placed in the 'margins' of the traditional religious-oriented division between *costumbre*, Catholicism and Protestantism. This was, in retrospective, necessary, for it was only possible to understand this from the testimony of someone who was 'in-between' worlds. This second chapter will return to the 'centre' and introduce a *costumbrista* family, the Ramos patrilineage. Unlike Marcelino, who left *costumbre* and thus is 'unprotected' and in danger within the traditional cosmology (although he is protected within the Protestant one and its spiritual practices), the Ramos family keeps making offerings and prayers to their altars, thus being protected and on good terms with the ancestors and the earth gods. However, of course, things are not always easy and, just as happened with Marcelino, the *costumbrista* cosmos is, like life, a complex endeavour, filled with difficult tasks and perils.

Besides introducing the Ramos family, this chapter will make the ethnographic account of a very particular ritual that, to my notice, has not been described in detail by prior literature and is, as we will see, possibly an innovation on former cultural models. In Momostenango, and elsewhere in Guatemala, ritual specialists called *ajnawal mesa* ('He of the spirit of the table') perform 'seances at midnight in totally darkened rooms, during which the ancestors and sacred mountains (*mundos*) arrive and speak through them' (Tedlock 1992: 74). These sessions, which usually take places at tables and in some cases incorporate theatrical elements like the rattle of chains, and blankets covering the performer, seem to have incorporated earlier patterns of shamanistic possession with the influence of nineteenth century spiritualism (cfr. Cook 1981: 186). However, the ritual described in this chapter, while being performed by a young *ajnawal mesa* who makes the *mundo* spirits speak through him, is rather different from its reported forms and incorporates other elements, including the mediation of the patron saint of the Ramos patrilineage, a peculiar image called San Antonio Mundo, the saint of the Mundo himself (Cook 1981: 64–75). The ritual is performed, essentially, to inform the family whether the ancestors and the *mundo* spirits are pleased or displeased with them, and to ascertain possible risks in their daily, private endeavours.

Similarly, this chapter will form a diptych of sorts with the next one, which is concentrated on a more public cult, that of Apóstol Santiago, the patron saint of the town. Both cults could be said to encompass the private and the public aspect of *costumbre* in regards to the Ramos family. However, as we will see, these rituals, which are heavily Catholic in many of their elements, incorporate the features of the paradoxical cosmology that I have outlined in the former chapter, features that, in a way, transcend or 'cut through' Western religious categories and give a particular, complex shape to K'iche' people's lives. In any case, it is necessary to start with the ethnographic account, which will go back to my arrival in town, when I was introduced to the Ramos family (Figure 2.1).

Virgilio

Shortly after my arrival in the town in late July 2017, which took place during the celebrations made in honour or the patron saint of the town, Santiago, I was introduced to Virgilio Ramos Ajtún,[1] a man of notorious energy and vigour of almost sixty, head of a *costumbrista* K'iche' household, thanks to the good offices of a common friend of both, the anthropologist Canek Estrada Peña.[2] After a brief interview with Virgilio, his wife Susana and his eldest son, Rudy, who is also a calendrical diviner as well as an *ajnawal mesa* 'spiritualist', and the most likely to succeed Virgilio as head of the patrilineage, the attention of all of us was directed towards the celebration that was taking place that week: the patronal fair (*feria patronal*), and its religious processions. Indeed, being a devout catholic, Virgilio is always involved in the *feria* and in the worship of Santiago in one way or another. As one of the sponsors of the local sodality group or *cofradía*,[3] he dons his best suit in the day of the procession

Figure 2.1. Left: Virgilio Ramos Ajtún, head of the Ramos family, poses with the patron saint of the family, San Antonio Mundo, inside his home. Right: Susana Argueta Xiloj, Virgilio's wife, at the *feria* of Santiago. © Alonso Zamora Corona.

and follows Santiago all day, while his family enjoys the *feria*. Thus, my first days in the field consisted mainly of following Virgilio around, watching him converse with other *pasados* and members of the *cofradía*, or accompanying his family to witness the participation of their grandchildren in a school festival associated to the *feria*. Rudy also had the kindness to take me to the main altars in Momostenango while explaining more details regarding *costumbre*. From the first days of my stay, a sense of indebtedness arose: Virgilio had, for example, helped me to find accommodation in the town centre, and presented me to some of his friends who were *principales*, making sure that nothing bad would happen to me as a recent arrival in town.

Almost at the end of the week of the *feria*, something happened that confirmed to me that this idea of being protected or 'recommended' (*recomendado*) by Virgilio was not only a matter of mere hospitality in the Western sense. After a long day accompanying Virgilio and watching the *feria*, night was falling. I took a moment to contemplate Santiago, carried by the *cofrades*, getting lost in the distance on his way to the Calvario Church at the graveyard of Momostenango, preceded by women carrying incense burners and followed in its trail by musicians, as well as by the sponsors of the *cofradía*, with Virgilio. Suddenly, a man pulled my arm. He introduced himself as Casimiro, a friend of Virgilio. He had heard about my interest in K'iche' 'customs' and 'traditions', the words I would say to excuse introducing myself with the meaningless word 'anthropologist'; it was obvious that word about strangers spreads fast in Momostenango. He told me: 'Do you want to see the *quemadero*?' referring to the altars where people would burn offerings of incense to the *dioses mundo*, the divinities of the earth. I did not refuse. He took me to the nearby Pak'lom hill, the main ceremonial centre of the town. Night had fallen by then, and the artificial lights of the street were still off, so our only source of light at the moment were the offerings that an old woman was burning.

Casimiro pointed at one of the many shrines at the place, and he said, 'Here I burn offerings' (*yo aquí quemo*), the usual expression in the town to say that one is a *chuchqajaw*. However, he added a little confidence to this, telling me:

> I haven't performed as much *costumbre* as Virgilio has. The difference between me and him is that I can burn offerings here, but he can burn offerings in many other places. He can make offerings on behalf of others, on behalf of you, for example. He can intercede for you before the earth, just as Santiago intercedes on behalf of us before heaven. He can make you successful, he can carry you, he has a gift [*tiene don*]. For he is before the *Santo Mundo* what Santiago is before heaven.[4] I know about your studies in England, know that you will owe everything to him. And, even if you want to, you will never repay him.

The parallel that Casimiro made was clear. Santiago, the saint of the town, is an intercessor on behalf of the whole town of Momostenango before the powers of Heaven or *Dios Cielo*. However, as a lineage head, Virgilio was somehow an intercessor on behalf of his family and clients before the *Santo Mundo* or 'Holy Earth'.[5] According to Casimiro, those people that Virgilio supported, as well as the shrines at which he had the privilege of performing rituals, were called *ekomal*, his 'burden'; thus, an unbreakable link between ritual and social obligations existed. He was not only responsible for the wellbeing of his own family, but also of his clients, and keeping this wellbeing is the foremost obligation of a *chuchqajaw*, as well as the foremost source of risk for their profession.

Besides this clear proof that performing offerings was not really a matter of curing people from 'shamanic' illnesses but also, and more crucially, a matter of sustaining their social efforts, from the very first moment I arrived in the town, I would come to be considered as something that

Virgilio had to care about in a sense, although he never really told me that directly. Nevertheless, that which Casimiro hinted at with the words *él te carga*, 'he carries you', was something very real in our personal relationship. Virgilio always showed a strong preoccupation with my person, even going so far as to help me find a place to rent in the town centre and presenting me to many people, something indispensable if one wants to wander the roads of the rural cantons of the town at night. Thus, my presence was not that of a neutral observer or a person who just could 'hang out' before the family life of Virgilio. I had to be placed into a certain position before both the family and the ever-looming ancestors and earth gods, the cosmological place that I presented in my last chapter. For example, when Virgilio and other *chuchqajawib'* performed ceremonies for others in my presence, they would always introduce me before the unseen powers they were addressing, be them dead ancestors or the *Santo Mundo* divinities, by offering them the same personal details I gave them when I presented myself. Asked about this, they would say that their ancestors were to be informed of my presence to proceed with any ceremony. Eventually, Virgilio would come to develop an interest in the very existence of my research; he considered it something that he had also produced or protected, which of course was the truth. But, furthermore, my research (and this book, too) was something to be placed within the 'Holy Earth', along with myself. However, at the beginning, I did not understand how exactly the idea of 'protection' worked. As the reader already knows, it was only later that I understood more.

The Ramos Family

Virgilio Ramos Ajtún and his family, members of a branch or line (*xeteil*) of the Ramos patrilineage, live in *caserío* Chonimatux, a small hamlet relatively close to the town centre that belongs to Barrio Santa Ana, one of the four main wards of Momostenango. According to Virgilio, many of the members of the patrilineages in the vicinity have ceased to actively perform *costumbre*, either because of conversions to Evangelic Protestantism or simply because of the death without succession of former *chuchqajawib'*, leaving him as one of the last two or three active *ajq'iij* or divining priests in the area.[6] Chonimatux has a school, an *auxiliatura* or local administration building, which are very near each other, as well as a Catholic church. Owing to recent conversions, there are now some Evangelic temples that, in contrast to the bigger Catholic church, are little more than small houses on the land of some people. The Catholic patron saint of the community is Santiago. In fact, Virgilio is the *ajq'iij* or shaman-priest of the local *alcaldía*, and he burns offerings on behalf of the *corporación* or government of the local *alcalde*. He is also a *pasado*, for he was once the local *alcalde*, which makes him a member of the council of *principales* who elect the new *alcalde* each year. Moreover, his capacity to perform ceremonies on behalf of the community and on behalf of other families besides his own (ceremonies like betrothals, weddings, funerals and protective rituals) make him, in the eyes of other families, a *k'amal b'e* ('road guide'), a leader for the community in both the spiritual and political sense. Thus, he is well considered as an important voice at the local level, often performing negotiations between different patrilineages.

Virgilio is married to Susana, mother of three children, all now married: Rudy, Neri and Joel. The Ramos brothers have multiple children too, and family life is usually very busy, since Virgilio's and Susana's sisters sporadically join in in family meetings, as do their nephews and their cousins. However, the Ramos Ajtún family lives a somewhat distanced life from other members of the Ramos patrilineage itself: while Virgilio is recognized as a reliable ritual specialist among *costumbristas* in the vicinity, some relatives of his patrilineage do not practise *costumbre* anymore, and thus are partly estranged from him. Thus, his patrilineage is split, and Virgilio's current main personal preoccupation is trying to rebuild a more united family by keeping a closer relationship between him and his sons than that which he and his brothers had with his father. As he told me, for him it was important that his family was in harmony, free from gossip, especially since he, as a *k'amal b'e*, was to be an example for the community of the prestige of *costumbre*; thus, any of his failings would imply a further loss of prestige for *costumbre* itself. The members of the Ramos family perform different activities: farming (maize, avocados, peaches), raising pigs and poultry, weaving ponchos, small-scale commerce, as well as taking temporary jobs in the town centre.

Truth be told, the life of Virgilio and his family was not 'dramatic' in any sense, despite the obvious economic hardships that Guatemala as a whole experiences today.[7] Most of my visits took place in a relaxed atmosphere, or perhaps felt as interruptions on the busy flow of the Ramos' lives, in which I took part: helping at the sowing and planting of maize, clearing the terrain for the construction of a new house, going to the end-of-year dancing soirée at the school, or enjoying a bath in the hot-springs of the mountains. Of course, setbacks and problems were not absent. For example, violent winds were prevalent in the year of my arrival, ruining crops for many; the leaves of some of the maize crops on the Ramos' land became infected and the affected plants withered; the market of wool blankets was undergoing a deep crisis, and thus Virgilio's

Figure 2.2. Left: Virgilio works his land. Right: Virgilio poses with a finished wool blanket with a *tinamit* or 'town' pattern (notice the little townspeople and a couple of quetzals in the centre). © Alonso Zamora Corona.

loom was not as busy as in the old days, just to name a few. Setting store in the town centre is also not easy nor cheap, and sales were sometimes slow for the family's pharmacy business. Likewise, the economic pressures of Virgilio's participation in the local *auxiliatura*, which required unpaid work (*patan*) from him and his sons to expand and consolidate public services in Chonimatux (like running water), were also felt, as well as those deriving from Virgilio's participation in the Catholic *cofradía* of Santiago, were actually rather high.

One day, Virgilio showed me his main project: to build another Catholic church in Chonimatux or nearby, to counter the rising influence of Protestantism. Indeed, besides supporting his family, the main preoccupation of Virgilio was to make *costumbre* stronger, and to do this he wanted to counter the recent decline of Catholicism, which has lost terrain to Evangelical churches. Indeed, besides being a prominent *k'amal b'e*, Virgilio is also strongly Catholic, even going as far as being a member of Acción Católica, a group known for its history of conflict with *costumbre* (cfr. Tedlock 1992: 40–43; Carmack 1995: 233–6); in the present, although the Catholic Church has never lost its intentions to purge local Catholicism from Indigenous elements, its power has greatly diminished before both Protestantism and Mormonism. Thus, some people like Virgilio strongly support Catholicism, since, as he explained to me, 'Catholicism doesn't forbid *costumbre* anymore', while Mormons and Protestants persecute *costumbre*, identifying it with 'idolatry', as I myself found in the speech of recent converts. As a part of Virgilio's involvement in Acción Católica, he is affiliated to a cooperative, Cooperativa Salcajá, a financial cooperative society founded by members of Acción Católica to develop projects for their affiliated families. Within the cooperative, Virgilio's main purpose was to raise money to build the Catholic temple. The problem is that, at the time, the project required an enormous sum of money to be completed, so saving enough to begin its realization was Virgilio's main long-term preoccupation.

While things went on with normalcy during my stay, the truth is that all the activities and plans described in the earlier paragraphs implied a considerable investment of money, time and work. The risk of failure at some of them was considerable, too. As head of the patrilineage, Virgilio's responsibility was to ensure that all these activities were successful, and, of course, the success of these activities was the actual motivation of the rituals carried on by him and his son Rudy on behalf of the whole family. As mentioned, these rituals took place at the *awas rech alaxik*, the ancestral altars of the patrilineage, scattered across his lands and fields. I have already introduced the meaning of the altars, but exposing how the Ramos family explained it to me is important here.

The Altars of the Ramos Family

The sacred altars of the K'iche' are by now familiar to the reader as the centre of K'iche' cosmology (Figure 2.3.). As *costumbristas*, the Ramos family also have these altars, which are called, as mentioned before, *winel*, *meb'il* and *warab'al ja*. Their *winel* is a big structure made out of stone and broken fragments of ceramic and clay (*xot*), which lies in the middle of one of their maize fields. They also have a smaller *warab'al ja* shrine near a river, and another altar called *mesa* ('table'). The *mesa* is considered to be the core (*matriz*) of the other altars, the point where the other altars converge; it is considered to be of a higher standing than the others, and particularly strong and dangerous. The unity of all these altars is called *awas rech alaxik*, or 'taboo of the patrilineage'. The ceremonies performed at them involve the cleaning of ritual spaces that are covered with pine needles, and the burning of offerings of native incense, candles, wood and foodstuff, while prayers directed to the ancestors are recited.

The functions attributed to each type of altar by the Ramos are a bit different from the classic scheme reported by Carmack and Tedlock: the *winel* is considered by them to be both the agricultural and family altar, the *warab'al ja* is only for the protection of travellers, while the *meb'il*,

Figure 2.3. Sacred shrines. Upper left: *winel* type of altar. Upper right: abandoned *warab'al ja* altar. Bottom: *meb'il* family stone shaped like a rabbit. © Alonso Zamora Corona

unlike the others, is manifold: it is a portable stone with the shape of a rabbit, together with all the precious stones found by the family in the mountains, which are considered to hold the presence of beneficent mountain spirits and, after a consecration ceremony (called *saturación*), become amulets of good fortune for the increase of money and human life. They can be placed on the inside of the *winel* to be kept safe.[8]

As mentioned before, I was offered many explanations about these structures from time to time; for example, after helping Virgilio and his sons clear the foundations of a new house, although I was not capable of understanding them. I was offered a more detailed explanation a month after the sowing of maize in the Ramos' *milpa* or cornfield. The young maize stalks need to be reinforced by making a small pile of earth with the hoe, so the stalks will not bend under strong winds and storms, a work that is called *calzado*. The whole family takes part in this work, as well as in the sowing and eventual harvesting; the day favoured for this agricultural work is Q'anil, the day sacred to the *milpa*, and this day is used to perform ceremonies at the *winel* altar, which has agricultural connotations. Commenting on the meaning of these structures, Virgilio offered me at first a vague explanation:

> *Winel* is the heart of our family. We are offered there since we are born, and when my grandson was born the altar was thanked for the arrival of another member of the family. Each family must have one, and that is what we call *awas* … This altar is called the *awas Q'anil* (altar of the day Q'anil) and here we ask for corn, water, we make ceremonies when women are pregnant, and also when a baby is born.

Then Virgilio expanded on the meaning of the whole array of altars, insisting that the *mesa* is especially dangerous (*delicado*),[9] and that they must be always fed with offerings so the family is 'up to date' with their ritual obligations. He said that offerings must be made at them with a proper ritual procedure, including abstinence from sexual contact the night before a ceremony is carried out. As is the case with the San Antonio Mundo Hill, this *awas* also has a guardian spirit, a *nawal*, which sometimes takes the form of a snake and other wild animals. The offerings are made so the *nawal* of the altar will not 'bite' people when neglected:

> [The altar] is guarded by a snake, in K'iche' we call it *nawal*, in Spanish it is called *santoral*[10] … the altar, when it is not cared for, bites (*kationik*), if you don't know how to make an offering, it kills, it kills you … My father placed my name at the *winel*, since he was a follower of the way of the ancient ones, he placed it. When my grandson was born, we gave notice here, for this is the place where the snake rises, and the snake enters in the corral, the coyote enters, the wildcat, they eat the animals, but if we give the offerings, we are protected …

The imagery used by Virgilio in talking about the *nawal* or spirit of the altar 'attacking' animals in a little animal pen or corral was at first obscure to me; it seemed only like a vague metaphor of punishment. It was also unclear whether he was referring to his domestic animals or his children being punished by these spirits. Other researchers came across similar images, like Garrett Cook, whose informants told him that the altars were made so wild animals would not come to the lands of people and attack their livestock, in a literal fashion (Cook 2000: 99). But, by now, the reader knows these animals are not conventional or common animals, but *awajmundo* or spirits of the earth gods. Therefore, the animals that Virgilio said that 'rise' from the altar are in fact underworld forces that, hostile for the family out of their possible neglect of the ancestors, can attack them. Later, Rudy confirmed to me that this idea was correct. When asked later whether the altars 'rise' (*se levantan*) and 'bark and bite' (*ladran y muerden*), as Marcelino told me, Rudy replied:

> Yes, the *awas* rises. It is sometimes a person, sometimes an animal. Sometimes it announces when you are failing [in your ritual obligations towards] it. It arrives, it turns into a snake. I was working on a 'case' this week, because during the night a family arrived to have dinner, the family head was sitting in the kitchen, his wife was bringing some firewood and the kids went to bed, and out of nowhere a snake was coiling in the kitchen. There was nothing there before, the man was watching the door. In the blink of an eye, a snake arrived. It was coiling. This was because the altar came and warned them: 'You are failing, defend yourselves, feed me' (*defiéndanse, me están fallando, dénme de comer*). My blood vibrated and told me that it was the *awas*, the *awas* is located on the navel. When you fail, it bites you ...

A bit of commentary is necessary here. Besides caring for their own family, both Virgilio and Rudy perform ceremonies on behalf of other families, and also carry out divination in order to ascertain where the problems of other people come from. These are called 'cases', as if they were 'detectives', in a way. For example, as explained, in this 'case' a coiling snake appeared in the house of another family, which was recognised by them as an omen, but not of which kind. Rudy, who has the 'gift of the blood' explained in the introductory chapter, can ascertain the true reason behind this appearance. By merely telling him about this experience, he can feel in his blood the identity of this *mundo* being and, furthermore, he can hear its voice, which is an actual voice of an altar, which can talk and communicate with diviners, as Marcelino explained to me, in the form of sudden voices, dreams or omens. These voices can be hostile, but they can also be pleasing, since altars are not only the source of the hostility of the underworld, but the sources of all earthly success. As Rudy explained me:

> Let's say one desires a job, one asks for it at the altar of the ancestors, all right, it 'opens the doors' for you, and from nowhere you are employed. Now, let's say I want to marry, and from nowhere the ancestors grant this to you too. One says: I want children, and they grant children. Everything in life is there, it is registered. When one will die, he goes there with the ancestors. For they say in dreams: this person will die.

This notion of registry or 'writing' (*tz'iib'am*, 'what is written') is vital to the K'iche' idea of 'destiny', and evokes the affinity of Maya cultures towards written records.[11] On one occasion, for example, Rudy and I were talking about the rituals that must be performed on behalf of a dead person of importance. When a *chuchqajaw* dies, his successor must perform a ceremony on the Ajpu day, the day in the Maya calendar that is dedicated to the dead ancestors. Rudy said: 'At the day Ajpu, we ask on behalf of the dead. One must wait for the day thirteen Ajpu, the optimal day to be accepted among the ancestors. They are gods, the dead are gods. They remain among us and speak through their altars. They punish those who they disapprove of.' The K'iche' term *komon* denotes these divine dead ancestors conceived as a unity: lists of their names are a common part of the paraphernalia of a *k'amal b'e*.

As guides of people's conduct, ancestors both grant punishments and blessings to the living members of their lineages. They are said to live below the earth or inside the summit of ancestral mountains; this is why it is said of them *saq kiwi, saq kijolom* ('white are your hair and your heads') a term that denotes the white clouds at the summit of the mountains (Schulze-Jena 1933: 20). Furthermore, Virgilio insisted that the fortune of his patrilineage and his capacity to hold agricultural land was granted by the intercession of his ancestors at this altar: 'Every plot of land that I want to have, I can have it, only with the permission of the altar;' therefore, part of the danger associated with the agency of this place was precisely because any expansion on social and economic power must be accompanied by corresponding ceremonies of thanks-giving at these altars. Another informant told me that he would increase the size of the altar depending on his economic and social successes; some others would say that further pieces of stone and clay would be added for each new member of the patrilineage, and so on.

Thus, just as social success is attributed to the intercession of the ancestors, which form a sort of cosmological society, social failure is attributed to the attack of the animal spirits of the underworld and the anger of the ancestors. As Rudy told me:

> One would say, why is this happening? It is because of our debt with the ancestors, one makes an offering and the thing stops. If not, after you die, the altar insists, insists so the children are up to date with the offerings. If not, *it can take the whole family with it*. The altar is that way, the altar is fear-some, it makes you poor, it puts you into nakedness, into madness, for what won't it do to you!

All these testimonies confirm the ambivalent nature of the altar. While being the foundation of family and social life and the seat of the beneficent power of the ancestors visualized as earth gods, the altar is linked by a powerful, deep cosmology to the destructive, primordial forces of the earth. Virgilio even told me that sometimes the *awajmundo* (jaguars, monkeys, deer) can be seen near the neglected *awas* emanating heat through their *nawal*.

This is all well known to the reader by now; and the reader knows my hypotheses regarding the 'topology' suggested by such a place. I will now present a concrete example of its presence in a ceremony of protection associated with the *awas* of the Ramos family, to show how all the former economic and daily-life preoccupations I have alluded to are engaged in a private family ceremony in which the whole patrilineage participates, and in which I had the privilege to participate.

San Antonio Mundo and a Ritual Seance

The ceremony I will describe is the festivity of the patron saint of the Ramos family, San Antonio Mundo, called by some people of the town 'the saint of the earth' (*el santo del mundo*). A complex cult of native saints exists in Momostenango, which has been extensively treated by Garrett Cook (2000). Among them, two are the foremost: Santiago, the patron of the town, and San Antonio Mundo. Santiago, of whom we will speak more in the next chapter, is only on surface the Catholic Saint James; likewise, San Antonio Mundo is partly inspired by San Antonio de Padua, the Portuguese saint that has the privilege of carrying baby Jesus in his iconographical representation (Hugh Farmer 2004). As Garrett Cook explains, Santiago was invested as the 'captain' or warlord of the Holy Earth by the old *principales*, but San Antonio would be considered as the true Santo del Mundo, the saint that represents the Holy Earth itself, where ancestors and earth divinities live. Owing to the fact that San Antonio de Padua is the patron saint of animals, he was considered the protector of cattle and domestic animals, and thus became San Antonio Mundo (Cook 2000: 64–65). As we have seen regarding the relationship between *nawal*, the animal 'co-essence' of people, and *santoral*, this is not a coincidence. Most curiously, the San Antonio image adored at Momostenango's main church only consists of the *niño* or baby Jesus part, without saint Anthony himself!

Just like Santiago, San Antonio is taken on local pilgrimages to cattle-raising communities in August. I have witnessed the cult status given to the image by the *chuchqajaw rech tinamit*, which performed a ceremony on behalf of the *auxiliatura* of Aldea Santa Ana, his own community. However, I will touch upon this particular topic in Chapter 6. For now, I will refer to another cult related to San Antonio Mundo, one that is more private in nature and, to my knowledge, previously unreported, and that parallels in some ways the *mesa seances*, although with a Catholic image rather than the *mundo* deities directly.

Besides giving cult status to their ancestors at their altars, some patricians are in possession of Catholic images that are said to have been found by their ancestors at mountains or underground. In fact, even the most powerful images of the town, like Santiago, have similar origin stories. For example, Santiago's image was found by one of the foremost lineages of the town, either the Vicentes or the Herreras, but eventually, as Momosteco lore affirms, the image would magically appear overnight at Momostenangos' town centre instead of the old town of *Ojer Tinamit* or Pueblo Viejo, and thus the image became the patron saint of the community instead of just a particular lineage. In a less public way, the Ramos family has a similar story about how they found their image of San Antonio Mundo. According to Rudy, the great-grandfather of the family was not from Momostenango, but from El Palmar, where he worked at the cane and coffee plantations of the coast:

> My great-grandfather worked at the cane fields, that's what the people before did for a living. One day, at lunch-time, he heard a voice calling him from the earth: 'Come, here I am.' He was called many times daily, and sometimes when he passed by a mountain near the river bank, he tripped many times on the same stone. 'Why is this happening?' he said. So, he took his hoe, and started digging. Finally, he reached a stone, it seemed like nothing. But the stone guided him to an image, a San Antonio de Padua, and it is here still, it belongs to the family. So, he came, built his house, and everyone placed candles to honour him: my grandfather, my father and myself. He is not at our household, but in the original house where he placed it; it is not with us, but we are given permission to pay him cult [status]. He is the *santo ri alaxik*, the family saint, for he has a shrine too.

According to Virgilio, San Antonio protects the family, conserves the *milpa*, and intercedes on behalf of wealth, animals and businesses. The image is housed on the lands of Virgilio's cousin, although he does not partake of the cult anymore. It seems that neighbouring families would also have their own images: a patrilineage in the vicinity has an image of the Black Christ of Esquipulas, for example. Furthermore, the saint is not only part of the origins of the patrilineage, but he receives actual cult status; the image is used during a divining ceremony that is performed each year on the day of Saint Anthony, a ceremony in which all the members of the family take part. The main objective of this ritual is to obtain an omen through the image, so the family knows what is going to happen during the year; sometimes the omen spells good, sometimes misfortune. As Rudy told me about a past ceremony:

> When his day arrives, we will venerate him. We bring him, we pray to him, and say to him, 'Take care of our animals, our road.' We make a circle at the patio with all the family, which we venerate, some of my cousins, my uncle. Three of four years before, the saint came and told me, 'Someone in the family is going to die.' I was concentrating, and he told me that. I almost

> didn't feel when he told me, but six months later, my cousin died. And when he died, my relatives arrived to tell me, 'You knew it was going to happen, you didn't say which one of us, but you knew he was going to die.' And yes, if I told them whom, they would not believe me, but it is a gift from the Mundo, it grants an omen. Last year, all went fine, but now we will make another ceremony to see what will come to happen next year. That is the *santoral* of the family.

Rudy being an *ajnawal mesa* capable of transmitting the messages of the spirits of the *Santo Mundo*, this ritual inevitably remembers the seances described in the introduction of this chapter. While not a *costumbrista*, Abraham described to me the performance of an *ajnawal mesa*, which was commissioned by an acquaintance of himself on behalf of his *komon*, and which involved a spiritualist sitting on a table covered by blankets. It was performed in complete darkness and involved the rattling of chains. The *ajnawal mesa* began his performance with the following words:

> He said: 'We the gods have arrived, we who are gathered, the earth gods, the lords of the face of the world.' For it is believed that the dead are the earth gods, that the dead are gods, because they are *ajaw*, but not one, many, because they hear, they listen, they visit, we can ask everything for them, they are gods … The lords of the face of the world grant us things, they give health, wealth, fortune, they listen to us, they help us, they are everywhere, they can hear everywhere, they help people … They are ordered by their patrilineage, I cannot invite nor propitiate another group, but only my own lineage, my name is Ajiataz, so I would have to invoke the Ajiataz ancestors … This is what is called *komon*.

And indeed, on 13 June 2018, the day of Saint Anthony, the Ramos family gathered for a seance session of a similar kind. The celebration started with a family dinner with meat, beans, rice, chilli and tamales. After dining and talking about family issues (Virgilio complaining about his many work compromises, the young men of the family discussing soccer, the women discussing the future birth of another kid in the family), Virgilio took the image of San Antonio from its shrine at another neighbouring home (which was not on his land, but in his brother's land) and showed it to me. It was a full body image of Saint Anthony of Padua, carrying baby Jesus, cloaked in his Franciscan habit, with an oversized red rosary and a sombrero. The image also had a *meb'il* or precious stone associated with it, a small crystal encased in a rusty metal locket, impossible to open now, gleaming through a piercing in its container (Figure 2.4.).

Figure 2.4. San Antonio is venerated at a seance at one of the family's houses. © Alonso Zamora Corona.

The ceremony was not performed at the main household where we had dinner, but in a larger building near the house of Neri, Rudy's brother, a building sometimes used as a guesthouse for the family. The ritual was performed in a very large concrete room, where many wooden stools and benches were placed. San Antonio itself was placed over a cloth on the floor, before a baked clay tile (*xot*) with candles, as well as a broken pot where sugar and incense (*pom*) were placed. Thus, the ritual was not completely Catholic in tone, but more akin to the Indigenous ceremonies for the ancestors. It started with Virgilio addressing the image in a long prayer, asking for protection for all the family members as well as their jobs and businesses: the poncho loom, Neri's pharmacy, Rudy's newfound job at a *cooperativa*, the *cofradía* of Santiago, and so on. After this introduction, a pair of speakers were brought to the place, and marimba music was played.[12] We all started dancing then, moving slowly to the compass of the marimba.

Meanwhile, other members of the family were just chatting, laughing and being merry. The artificial lighting was shut off, and we were illuminated only by the candles. Rudy sat in the middle, before the image of San Antonio, and started a long shamanic chant. The prayer was mostly a very long concatenation of sacred beings: after invoking San Antonio as the 'shepherd of the *Santo Mundo*' and the family's ancestors or *komon*, Rudy addressed Santiago as 'captain of the *Santo Mundo*', and the souls of the neighbouring patrilineages, the *oxlajuj may*.[13] While repeating names and formulas, Rudy fell into a sort of deep trance in the dark. Finally, at some point, Rudy began to speak on behalf of the image. He murmured a prophecy to his father: 'There is danger, at the altar, sorrow is coming, but you are only going to witness it, only witness the tears, when it arrives …'

Rudy conveyed the information to Virgilio, who immediately started to show concern and deliberate which ceremonies were adequate to avoid this destiny. However, the family was not disheartened, and they kept dancing and being merry, despite Rudy's utterance. After the communication of this vision and the end of the dance on behalf of San Antonio, the members of the family were placed inside a circle of incense made by Virgilio. It was the 'corral' where the *nawal* of the patrilineage, their animal co-essences, were placed. When I entered the circle, one of the women of the patrilineage said, 'Alonso Ramos!' in a joking manner, and everybody laughed, me included. After that, we all passed our limbs over the fire, and were finally blessed by the image (Figure 2.5.). Finally, we went to Neri's house, where I eventually fell asleep. In the other room, Virgilio was making calendric counts to himself, trying to devise a day that would be adequate to avoid the bad fortune. He mentioned the day Keme or 'death' in the Maya calendar, possibly to avoid a mortal peril. A month later, while we were carrying some timber, Rudy would explain these omens to me. He said that the saint would again prophesize the death of someone. So, perhaps, that is why Virgilio had chosen the day Keme as appropriate to counter the omen. The other omen was a punishment for some failing of the family, but he would not specify which.

As Virgilio would later explain, the ceremony for San Antonio is an act to 'bring us together and make us a unit as a family before the *Mundo*'. And indeed, certainly the ceremony was an act of bringing together the family, for not only Virgilio, his wife and his sons and grandchildren would participate, but also some of the cousins that were close to the family, and even me. However, a deeper significance of the ceremony, associated with K'iche' cosmology, can also be discerned. As we have seen, the ancestors of the Momosteco families keep guard over the animal co-essence of their descendants, their *nawal*, in the *awas*, ensuring that nothing bad happens to them. This underworld space, called *korral*,

Figure 2.5. The ritual of San Antonio Mundo. © Alonso Zamora Corona

was represented by the circle of incense in which the family (and I) were placed at the end of the ritual. Punishment by the ancestors is exerted on the *nawal* co-essence of people, so the act of placing people inside the *korral* is a protection from the prospective punishment of the ancestors. In this, the ceremony is not too different from the protection rituals of the Tzotzil, registered by Vogt, where patients would enter decorated beds symbolic of the 'corral' where their *chanul* or companion animal is reintegrated (1969: 442–4). In the case of Virgilio and Rudy, the closing circle of incense was this act of 'closing the door'. However, this ceremony had not only the purpose of protecting the family, but also to peer into the future; Rudy would explain this to me in a way that is explicitly linked with the cosmology of animal souls.

Indeed, some time after the ceremony, I was talking with Rudy about it, as well as about the concept of *nawal*, which implies an animal soul. Rudy, of course, would explain the usual conception that is known in the local anthropological literature as *tonalismo*; that is, an animal soul that is linked to the destiny of the person.[14] However, Rudy added something particular regarding the ceremony. As he explained, the reason why he as a diviner can peer into the future is that he is able to hear San Antonio Mundo and speak to him through his animal soul or *nawal*. He would say to me, regarding his *nawal*: 'I'm an eagle. When I burn my offerings, my eyes heat up inside the *Santo Mundo*, and I can see far away.' This heat force is called *q'aq'il* (or *nawal* too) and it is the same force that animals and the altars themselves use to hunt and kill. Thus, San Antonio Mundo becomes for him a way to channel into the perspective of the ancestors and underground gods regarding the future of people, and visualize the possible harm that can come to them through the *awas*. According to recent interpretations on K'iche' ontology (MacKenzie 2016), this would imply more of a hierarchical ontological arrangement, where humans are seen as domestic animals of the ancestors, and nature is the material culture of divinities (Zamora Corona 2020).

Discussion

This chapter can be seen as a portrait of a K'iche' family, and the way in which their personal life and preoccupations often intersect with cosmological ideas and practices. The relationship between both of these facets of K'iche' life will be deepened in the following chapter, but there are some aspects of it that need to be discussed before moving forward. In general, it is necessary to assert that, besides being theories about 'spirits', 'souls', the world and the environment, Indigenous cosmologies are also social theories. For example, in my experience, and contrary to the expectations of Western practitioners of 'Maya spirituality', nobody approaches cosmological matters in Momostenango from a purely 'spiritual' starting point.[15] Instead, concrete life preoccupations are most of the time invoked as the real motivation for getting involved with rituals and cosmology.

Of course, by saying this, the ghost of functionalism, so thoroughly criticized in the context of Maya ethnology, seems to rear its head in an almost anachronistic fashion. Thus, the cosmology described by this work could be very much aligned with the ones described by the structural-functionalist anthropologists of the Chiapas school (Holland, Vogt, Cancian, Hermitte), since the idea of a 'corral of souls' is a feature of the cosmology of the Maya of the Highlands of Chiapas that is shared with those of Momostenango, and with other groups in the 'Mesoamerican' region.[16] Of course, no one considers themselves a follower of old-school structural-functionalism now, as Alan Barnard has remarked (2000: 73); however, in a defence of the 'old school', at least in the local context of Chiapas and Guatemala, the problems that classical theories tried to solve have not been sufficiently addressed by the regional post-functionalist ethnography, which simply looked elsewhere. Thus, while structural-functionalism and its (problematic) solutions to the relationship between social life and cosmology are rightly considered passé, in the end, the problem subsists and, even more, old theories were actually right in stating that social life is of crucial importance to the presence of cosmology in people's lives, even if they misinterpreted Indigenous cosmologies as being mere mirages of 'functional' preoccupations. This is not an assertion of anthropologists, but of informants themselves; for example, Virgilio explicitly mentioned to me that all his lands and his social standing depend on his relationship with the *Santo Mundo* in a very real sense.

As the reader has found during this chapter, people indeed express what we could call 'native functionalist' statements all the time, of the type 'I perform this ceremony in order to solve this (social) problem'; however, what I suggest is that we are not conceptualizing correctly the entanglement of the social and the cosmological in these affirmations. People do not perform rituals in order to solve ideal, cosmological situations (which would be the point of view of 'holistic' approaches), nor merely in order to 'justify' social actions (functionalist, sociological approaches), instead *people perform rituals in order to act socially in a cosmological fashion*. In fact, as we have seen in this chapter, these rituals are sometimes even more paradoxical. Usually, in the classic functionalist logic, we suppose that people only use cosmology in order to acquire prestige or solve a problem. But, for example, what Rudy was doing in the divination ritual was not trying to solve any problem, but trying to ascertain whether upcoming situations in the life of his family were to be considered as cosmological in nature. Thus, cosmology often reframes and recontextualizes concrete life situations, but this does not mean that cosmology is a mere ideological 'masking', as functionalists asserted. Instead, in this respect, it works in a similar way that social theories do in Western societies, and individual K'ich'e cosmologies are indeed 'cosmologies of the social' in the terms of Morten Axel Pederson (2011, 2017), cosmologies preoccupied with the possibilities and problems of concrete life, constantly reimagining the social, rather than merely being 'functionally' determined.

Finally, and in a way related to this problem, it is also necessary to talk about my presence in the ritual, an element that inevitably had some influence on its performance and outcome, inevitably recalling the idea that a perfectly objective ethnography is impossible, since the observer themselves changes the social situations they describe with their mere presence, problematizing the notions of subjectivity and objectivity (cfr. Hegelund 2005). At the start of this chapter's ethnographic account, I talked about how I was placed in a relationship of protection and even patronage in relationship to the Ramos family. Similarly, I have explained that, according to them, the ancestors have lists of names in the underworld, which state who is to be protected and who is not. This idea is crucial to the functioning of their cosmology, and, as I will explain later, lend these altars a sense of history and development. As it was explained to Barbara Tedlock: 'These shrines are like a book where everything – all births, marriages, deaths, successes and failures – is written down' (1992: 202). But the first, necessary, thing to 'enter' the altars is to be named within them. This is what Marcelino alluded to when he said, 'If I place you here, you are not going back to Mexico'. And, in a way, the meaning behind my many interlocutors insistently talking about me to their shrines was that: to introduce me to them, and place me within K'iche' cosmology (which is a crucial part of K'iche' society). Thus, my own place as observer was like that of the man in Escher's *Picture Gallery*, both inside and outside of any ritual and situation I partook of. The joking exclamation at the end of the ritual, while not to be taken very seriously (after all, it was not uttered by the shamanic performer, but by another member of the family) was, I think, also a somewhat joking way to introduce me to the *mundo* spirits, and associate my name with that of the family, reaffirming their conviction (which they expressed many times to me) that this book is also their work in a cosmological fashion, too.

Notes

1. Or 'Virgil', in English.
2. Estrada Peña is himself an anthropologist working with the K'iche' of Totonicapán, on which he has published several articles (2013, 2015).
3. In Catholicism, a sodality is a devotional community or confraternity. The most important work on sodalities in Momostenango is, as mentioned earlier, Garrett Cook's *Renewing the Maya World* (2000).
4. The earth itself, conceived as a cosmological dimension.
5. Later, I would realize that Casimiro's self-deprecating words in praise of his friend were a case of false modesty. He is too the *chuchqajaw* of a neighbouring clan, so naturally he also makes offerings on behalf of others and is well respected as a *k'amal b'e*, and holds the same kind of power he described through the example of his friend.
6. According to Virgilio, some of the prevalent *alaxik* in Chonimatux are Ramos, Chun, Ab'ac, Martínez, Ajka, Díaz, Zárate, Pelicó, Ixkoy, De León, Argueta, Xiloj and Cuyuch', with Chun, Ramos and Martínez being the more numerous.
7. According to the United Nations Development Programme, more than half of the population of the country lives in poverty; up to a 79 per cent of the poor in the country are Indigenous people (2009: vii).
8. In my first publication on the subject (cfr. Zamora Corona 2019), I followed the information of the Ramos. In the next one (2020), however, I followed the more traditional scheme where the *warab'al ja* is ancestral, the *winel* agricultural and the *meb'il* is financial, for there are calendrical reasons to support this interpretation as the more 'canonical' scheme, since the day-signs associated to them (Aj, Q'anil and Tz'ikin) are fixed, and have strong connotations to the idea of ancestors, agriculture and money, respectively. In truth, many variants within this scheme are found in the ethnographic literature (cfr. Cook 2000: 99 for another interpretation), and probably each patrilineage follows a particular variation.
9. This seemingly innocuous word is of great importance to K'iche' ideas. As defined by Garrett Cook: 'This Spanish term is generally used by Momostecans in the translation of Quich'e ideas about contact with the supernatural; it connotes the combined sensibility and danger associated with such undertakings' (2000: 267). The equivalent K'iche' terms are *itzel* ('powerful' or, sometimes 'evil') and *kaxlik* ('painful').
10. Or *sanctorale* in Latin, meaning the cycle of feasts dedicated to the saints in Catholic liturgy. The usage of this word to mean the hidden spirit or *nawal* of a thing is extraordinary, and it strengthens Pedro Pitarch's hypothesis on the equivalence between Catholic saints and *nahuals* in Indigenous Maya thought (2000).
11. A logosyllabic system of writing has been recognized as a crucial feature of Maya culture (Sharer and Traxler 2006: 125–52); however, the K'iche' writing system was probably closer in form to that of the Aztec (Garay 2014), which was both logosyllabic and pictographic (Zamora Corona 2022).
12. A keyboard percussion instrument that consists of a set of wooden bars struck by mallets. Of complex and uncertain origins, it has become central to the musical identity of Guatemala (cfr. Godínez Orantes 2015: 92–93; 319–21). Traditionally, in Momostenango, the veneration of the saints is always accompanied by marimba music and dancing, as I witnessed on many occasions.
13. 'A cohort of temporal groupings of the souls of the dead. A *may* may correspond to the dead of a generation or a century' (Cook 2000: 268). In Momostenango, the *may* corresponds to a period of twenty years (*junab'*), and it is also called *le'* ('row, line')

corresponding to the cycle called *k'atun* in Ancient Yucatan (cfr. Carmack 1981: 121). The *may* are said to conform to a group of thirteen, and each of them corresponds to the most prominent *alaxik* or patrilineages in the town and their territory or *parcialidades*, an ideal pattern to be found in both Classic and Colonial Maya culture (cfr. Rice 2013).

14. See Lamrani 2008, Martínez González 2016 and Monaghan 1998, for general anthropological and historical accounts on *tonalismo* and *nagualismo*.
15. For example, the occasional foreign practitioners of 'Maya Spirituality' that I met in Momostenango often dismissed the more concrete preoccupations that often motivate Indigenous cosmological practices (the ancestors granting health or income) as 'unspiritual'.
16. See Millán 2019 for a general reflection on the idea of 'corrals', and spiritual owners of animals and *naguales* in Mesoamericanist ethnology, and the possible influence of European cattle-raising practices on its development after the conquest.

3

A Future that Already Happened

Time Magic in K'iche' Rituals

Following the themes introduced in the last chapter, these pages will, on the surface, continue to present the reader with the classic anthropological topic of the search for prestige and power through social and ritual practices. Indeed, this chapter forms a diptych of sorts with the preceding one, going from the level of family endeavours and preoccupations to that of local politics and culture. Concretely, it will present the ethnography of one of the most important rituals that the Ramos family organizes: a local *feria* or celebration in honour of the itinerating image of Santiago, the patron saint of their community, Chonimatux, and of the whole town of Momostenango. This celebration, which implies a considerable expenditure in terms of money and time, is performed, as Rudy and Virgilio themselves told me, to consolidate the position of the *costumbristas* in Chonimatux, among other purposes.

However, while this ritual can be explained as a way in which the members of the *cofradía* or sodality group and the local political officers of Chonimatux participate in the wondrous power of Santiago (a warrior saint), another, deeper practice, unrelated to Catholic conceptions, takes place *before* the ritual. Indeed, the celebration of Santiago is not only rehearsed, prepared and performed by the members of the confraternity in a purely 'physical' manner. Beforehand, a series of preparatory rituals is performed, in which the events of the upcoming *feria* are described to the ancestors that will oversee them, compelling them to happen in the exact way in which the organizers intend them to, ensuring that they are properly executed in the *Santo Mundo* before they develop in our world.

To introduce the context necessary to understand these paradoxical preparatory rituals, which will be considered here as a peculiar form of 'time magic', I will first offer an overview on *cofradía* cults in Momostenango, which, as mentioned before, are the topic of an authoritative monograph by Garret Cook (2000). Then, the ethnographic account of both the preparatory rituals and the celebration of Santiago will be presented. Finally, a more theoretical analysis on the relationship between both will be developed, introducing a very complex topic: that of the anthropology of time (Gell 1992). Thus, I hope to show that, while calendrics and divination has occupied a central place in the ethnological literature on the K'iche' Maya, there are other processes and conception that also define the cosmological relationship of the K'iche' with time, running along the theoretical lines introduced in the former chapters; that is, self-reference, recursion, entanglements and paradoxes of hierarchy and inclusion.

The Cult of *Patrón* Santiago in Momostenango

It has been already explained elsewhere that both Santiago and San Antonio Mundo are Catholic saints only on the surface; in fact, they are more properly understood as being Indigenous saints. San Antonio Mundo has been introduced as the saint of the 'Holy Earth' and the protector of the souls of humans in the *Mundo*, being mainly concerned with family life and the harmony of smaller social units, as Garret Cook observed (2000: 74–75). In contrast, Santiago is a saint of the political sphere, a saint that embodies social and political power (Cook 2000: 97–98). The veneration of Santiago in Latin America was inherited from the conquistadors themselves, who considered him to be their protector. The devotion to Santiago was later adopted by many Indigenous communities, where he replaced meteorological and warrior gods (Capponi 2006), effectively changing from a conquistador saint to an Indigenous one. In Momostenango, Santiago has undergone further transformations: instead of wearing his traditional medieval armour, his image is dressed like a Guatemalan military cadet of the nineteenth century, wielding a sabre, covered in sumptuous ponchos and cloths, and being adorned with an elaborate feather crest on his head (Figure 3.1.). His own decapitated head, a product of his martyrdom, is displayed before him. He is paired with San Felipe, who is said to be his 'secretary', because it seems to be the case that Santiago does not speak Spanish, only K'iche' (Cook 2000: 95).

Figure 3.1. The image of Santiago at Momostenango's main church. © Alonso Zamora Corona

The Momosteco Santiago is somewhat of an 'earth god' or *mundo* himself or, more precisely, a sky god. He is called *nima ch'umil*, 'the great star', referring to Venus (Cook 2000: 28). Indeed, often, Rudy would invoke him by calling him 'captain of the holy earth' (*capitán del Santo Mundo*), for he is said to be the warrior that leads the army of the clouds and the stars.[1] Just as the fierce Venus gods of ancient Mesoamerica, who could 'spear' the spirit essences of crops, water, warriors and rulers when rising at certain calendrical dates (Milbrath 1999: 174–7; Nowotny 1976: 52), when angry, the Momosteco Santiago can 'trample' the *nawal* of crops and even that of rain with his horse, bringing droughts during the year, and is even capable of instigating political instability and quarrels within the hearts of men. According to Virgilio and Rudy, each year the leader of the town's *cofradía* receives a dream, indicating the augury for the town, which is manifest by the main colour of Santiago's clothes during his parade at the *feria patronal*: red clothes can mean political quarrels and deaths, black means rains, blue means lack of rains, multicoloured clothes are peace.[2]

The difference between the *costumbrista* Santiago and the Catholic Santiago is not ignored by the Catholic Church, either. Despite Santiago being kept in a chapel next to the altar of the main church of Momostenango, mainline Catholics in the town (like the few remaining Ladinos and some members of Acción Católica) reject his cult and actually hold a parade for the 'proper' image of James the Great or Santiago Apóstol, excluding all Indigenous elements from its iconography. This parade occurs the day after the one of the traditional Momosteco Santiago. I witnessed one of these parades, where loudspeakers explained in K'iche' the 'proper' biblical story of James the Great. However, these efforts are to no avail among Indigenous believers. Indeed, Rudy clarified to me that their Santiago is Santiago de los Caballeros ('Saint James of the Knights', the conquistador saint), not Santiago Apóstol (Saint James the Great, the disciple of Jesus). Thus, he claimed that 'Santiago is not in the Bible', a clever assertion, for this conquistador saint is both traditional and Catholic, but not completely biblical.

I have mentioned that Santiago is a political saint and, effectively, he is highly regarded by *costumbrista* politicians, including the former town mayor who (as I witnessed) liked to remind everyone (especially Protestants and Mormons) of the supremacy of the image during his political speeches in K'iche'. Santiago is considered by some to be none other than the *nawal* or 'co-essence' of Momostenango itself, and the relative stability and continued privileges of the town are attributed to him: some Neo-Maya practitioners in town told me that Santiago's calendric *nawal* was Kiej, the calendric day associated with political power. Virgilio, more of a traditional *costumbrista*, explicitly told me that 'the protection of the communities lies under the apostle Santiago, who protects the whole town'; thus, Santiago is also the protector of local *alcaldes* at Chonimatux, Virgilio's community.

Santiago has a native altar of his own at Puja'l Santiago, near the base of the Pak'lom Hill in downtown Momostenango, a place that his horse is said to have struck with its hooves, making a small water spring appear; this means that Santiago is not only revered at the Catholic church itself, but also at K'iche' shrines. Furthermore, Santiago is deeply connected to the town's history, for his image is associated to two of the founding lineages of the town, the Herrera and the Vicente families; in fact, the latter claim to have found Santiago inside a mountain, a claim that is repeated by local patrilineages regarding their particular patron saints, like San Antonio Mundo for the Ramos family. Thus, these native images or *wachib'al* come from the 'Holy Earth' itself and are its manifestations, and are considered as possessing a powerful agency that makes them 'delicate' (*delicadas*) or dangerous.[3] Rudy would tell me that the *wachib'al* of Santiago is capable of trampling people spiritually with his horse and even killing them, so it is not allowed that people fall asleep before him during the vigil nor are placed right in front of him during the local *feria*, as I myself witnessed, for when I positioned myself in front of the saint, one of the *cofrades* gently prodded my ribs to indicate me that this was forbidden. When Rudy was working with the *cofradía*, he would sometimes have dreams regarding Santiago's *wachib'al* coming to life before his eyes.[4]

In regards to the festivity that I will describe in this chapter, as mentioned in the preceding paragraphs, Garrett Cook (2000) has greatly elaborated on the cult of travelling saints; that is, the pilgrimages of Catholic images to faraway rural hamlets where they receive cult following by local *cofradías*, eventually reaching the communities of the patrilineages who claim traditional ownership of them. Thus, each year, in November, the image of Santiago visits San Vicente Buenabaj, where the Vicente lineage resides, as well as the canton known as Pueblo Viejo (*Ojer Tinamit*), where the Herrera lineage lives. As Cook remarks, this travelling cycle serves to claim and reconstitute the original founding relationships of the town, while reaffirming the authority of the town centre or *cabecera* over local lineages (2000: 64–65). However, before reaching their destination, these images spend some time in local hamlets and cantons, where their presence is also used to reaffirm the power of local authorities, blessing local icons as well as their devotees, including members of local *alcaldías*.

In the case of Chonimatux, Virgilio and other *pasados* from the locality are members of the 'Chonimatux Society of the Santa Ana Ward' or 'Santiaguito Society'. It was founded on 22 November 2008, the day of Santiago. Virgilio is not a current member of the society nowadays, but nonetheless continues to perform ritual functions and administrative functions for them, partly to still hold some influence in the local affairs, partly to lead Rudy into further political and ritual appointments and, most importantly, to retain contact with the *nawal* or spiritual essence of Santiago. At Chonimatux, an *ermita* or shrine is improvised each year at the house of a local family to do a vigil with candles (*velación*) and lay flowers in honour of the saint during the evening of 21 November, while the actual receiving of the image happens on the next day in the plaza near the building of the local Indigenous administration, called the *auxiliatura*. To accompany the celebration, a dance of merriment or *convite* is performed.[5]

Regarding the political importance of this public ritual, Virgilio told me that, since he is the calendric priest of Chonimatux, he burns offerings for the local *alcaldes* through the year: 'Santiago protects the incoming *alcalde*. The protection of the communities depends on Santiago; that is the reason why we venerate him, we take his icon to his altar and we present the offerings. Because he goes on his pilgrimage to another community, but we use that occasion to obtain our own blessings too.' Thus, many kinds of intentions and actors are present in the devotion to Santiago: devotees that want his blessing for themselves and their families, politicians that want to be protected by him, and calendrical priests who consolidate their role in *costumbre* by celebrating these rituals.

The Rituals of Anticipation

In preparation for the *feria*, Virgilio, Rudy and other men from the group perform a ceremony each month during nine months at Puja'l Santiago. Each of these rituals represents one of the years already elapsed since their commitment to the Santiaguito Society started, up to a maximum of nine.[6] I accompanied Rudy to one such ceremony, which he officiated and that took place on 13 August 2017, or 11 Ajmaq in the K'iche' *cholq'iij* calendar, a date propitious to saints. This timing in regards to the Indigenous calendar was important too, for the celebration of Santiago on 21 and 22 November would start on an Ajmaq day too; thus, a temporal connection regarding both dates in their relationship to the sacred calendar was established.

When I witnessed the ceremony, which took place some months after arriving in the town, I did not have a precise idea of what was taking place before me. My knowledge of K'iche' was bookish and limited at the time, and I had not sought help in working on the translations of the material that I had recorded until then. I could only understand that the ceremony was related to the petition of money and protection for the upcoming celebration, understanding some isolated words in K'iche' and, of course, what was said in Spanish. However, later, when working on the texts, something different came up. The ceremony was not only a petition for the success of the upcoming ritual or to obtain the money necessary to perform it, as was explained to me at first. Instead, in the ritual, Rudy addressed the dead founder of the Santiaguito Society, as well as his ancestors and Santiago himself, and invited them to the ceremony, describing all that would come to happen in a chronologically precise fashion, exhorting them to 'move' or compel the co-essences (*nawal*) of the people and even of the objects involved in the upcoming ritual, so the ceremony would be perfect. Of course, the ethnographic account presented here will take this into account, but I ignored it at the time.

Rudy began by burning offerings and recited a long prayer, in the style of many other K'iche' ceremonies I would witness during my stay at Momostenango (Figure 3.2.). After greeting the personified day lord 11 Ajmaq at the start of his prayer, he mentioned that the ritual was a *maltyoxib'al*, a thanksgiving ceremony, on behalf of all the members of the Santiaguito Society, especially naming the main appointees of it, the secretary (*secretario*), the treasurer (*tesorero*) and the president (*presidente*), as well as the members of the *vasalla*, the local dance company.[7] Rudy then mentioned that he had made a commitment to dance at the feast of Santiago during nine years, and implored for financial means both to be able to participate in this performance and be successful in it, for the members of the group had to raise Q30,000 GTQ (Guatemalan

Figure 3.2. Rudy and his companions burning offerings at Puja'l Santiago. © Alonso Zamora Corona

quetzales, ca. £2,970 at that time, £3,890 in 2025) throughout the year. Thus, he would implore his *meb'il* altar or the 'heart of money' (*k'ux ri pwak*) on behalf of his own capital, and also on behalf of the businesses of the members of the society in the town and even in Guatemala City, while mentioning many places in the sacred K'iche' geography during his prayer.

After this preliminary invocation, Rudy greeted the dead ancestors, collectively known as *oxlaju'j may*, both in their spiritual essence (*tew kaqiq'* 'cold and wind') and in their earthly remains (*alaq much'ulik b'aq, alaq much'ulik ulew* 'heaps of bones and dust'). He then counted thirteen *may*, the sacred period mentioned in the last chapter. He then presented his offering to the spirit of the founder of the Santiaguito Society, as well as the other deceased members of the group (*komon sociedad*) and the dead *alcaldes* of the locality, regarding them as intercessors before the ancestors and the earth gods, asking them to ensure that there would be no problems during the celebration. He then began describing the upcoming ceremony. First, he asked the ancestors to 'raise and bring' (*che' yekak'uloq, che' k'ama'uloq*) the spirits (*nawal*) of people involved: the absent members of the *sociedad* and the prospective guests, even naming some of them in person, so they would attend the celebration, their spirits being moved by their ancestors. Then, he would invoke the calendrical spirit of the two days of the K'iche' calendar on which the celebration would take place (6 Ajmaq and 7 No'j), and start informing Santiago and Felipe of the details and chronology of the event.

While describing the ritual, Rudy asked the ancestors not only to bring people 'from the four directions' to the celebration, but also compelled the spirit or *nawal* of the things that were going to be used in the celebration; he compelled the icon (*wachib'al*) of Santiago to dance, the money to keep flowing, the musical instruments to gather and seduce the people, the fireworks to be properly set, and even compelled the *nawal* of the alcohol to settle on people's feet and stomachs rather than on their heads, so it would not cloud people's judgement, but give them joy and a good time. He finished his prayer with a general petition to God, the ancestors and the saints:

> We ask for you who are in cold, in wind, in clouds and mist, Santiago and San Felipe, that they [the people] will be glad to join us. Before the visage of the ancestors, we ask that our petition and our prayer shine, that they gleam, that our words and our conversation be fragrant, that they reach the cold and the wind.

Thus, through Rudy's preparatory prayer, we can see the true powers working behind the stage of Rudy's and Virgilio's future performance.

As an *ajq'iij*, Rudy would mobilize in his prayer a number of cosmological beings involved in the prospective ritual to ensure that everything would be done down to the last detail, in a chronologically appropriate fashion. Thus, in a way, he compelled the ancestors to oversee and even move the spirits of the things and the people that would partake in the upcoming ritual, but in the ancestral dimension of the *Santo Mundo*. But a deeper analysis of this process will be made at the end of this chapter. Now, I will present the ethnographic description of the second ritual, or the celebration that was prepared in the ceremony described in the preceding paragraphs.

The Reception of the Image

Mid-November is one of the busiest times of the year for Virgilio and Rudy, for they not only have to perform ritually as *chuchqajawib'* in the feast of Santiago, but also ensure that the whole community is entertained with two days of concerts, feasting, dancing and socializing organized by them and the other members of the Santiaguito Society. As mentioned, in these events, both political and ritual motivations are intermixed. Before the wake (*velación*), Virgilio directed a small dance with Santiago's icon with the members of the Sociedad, as it is customary with the images of many saints in Momostenango. He danced to the compass of a marimba while incensing Santiago's image. Afterwards, he and his companions placed the icon of Santiago on his altar, and celebrated a small Catholic mass with Rudy. In this ceremony, they incensed Santiago's icon and blessed many people by placing a ritual bundle of flowers (*b'otz'oj kotzij*) over their heads. This bundle is in fact a symbol of political and ritual power among the K'iche', for both the local *alcalde* and the *chuchqajaw rech tinamit* have their own. Finally, he placed the *alcaldes'* baton (*vara*), another crucial insignia of authority, before Santiago, and suffumigated both the *vara* and the outgoing *alcaldes* of the canton while chanting prayers. The resulting image was one of deep overlap of the ritual and civil functions: Virgilio, a local *chuchqajaw* but also a local power broker, incensing both the local icon of Santiago and the local authorities in office at that time, as well as the insignia of their rank.

Virgilio would spend the evening discussing the organization of the Santiaguito Society and issues of the local *auxiliatura* with former *pasados*, as well as ensuring that the celebration was carried on properly. However, equally as important as the ritual aspect was the festive aspect. While the devotees of Santiago kept praying in a small chapel that had been improvised in the building, most of the people stayed for the loud concert that would follow during the night, performed by 'Ronald y sus bravos' a *cumbia* band that performed Mexican music while people were drinking beers of the local Gallo and Ice brands, as well as eating the *tayuyos* (tamales with bean sauce) and *pepian* (beef in red chilli and tomato sauce accompanied by white rice) offered by the members of the society.

After a night of dancing with the community, the family and I went back to their home, since at 4am everyone had to be awake again to prepare for the ceremonies of the following day. On the road home, Rudy started to jokingly quarrel with his brothers. He boasted that, by being an *ajq'iij* and future *chuchqajaw*, and by participating in the communal activities, he would get access to political authority, and 'even the keys to the soccer field', so they had better not upset him or they would not be able to play in the future. Later, they talked about those local families who did not participate in the *feria*, saying that 'sooner or later, there will be authority over them'. Rudy then talked to me a bit about his role as a dancer for Santiago at the next day's *convite*, and then he departed to his own home with his wife. We walked home, and I stayed at a guest-room at Neri's house.

The next morning, before the *feria*, Rudy and Virgilio woke me up at 4am. It was time to perform *costumbre* on behalf of the local icon of Santiago, before the image of Santiago arrived at the *caserío*. We took the path from Chonimatux to an altar near the main road to the town centre, *Ventana Mundo*, 'the window of the underworld', from which the town can be seen, but especially the cemetery. We walked in the dark, while other companions of Virgilio's group joined us, discussing affairs concerning the worship of the *Santo Mundo*, counting the days for the next ritual offering. Meanwhile, Virgilio explained to me that it is not necessary to be a *k'amal b'e* to obtain an appointment at the *auxiliatura*, but it makes it easier, for ritual service is also a service to the community and, certainly, as I could corroborate it during the ceremony, political authorities are helped by ritual ones.

We arrived before sunrise, and the rest of the members of the group were waiting. Virgilio placed Santiago's icon next to the altar, while the rest of us prepared the offerings of small pellets of incense, candles and big offerings of *q'ol* or pellets of pine resin, placing them over pine needles, while the altar was being adorned by flowers. A cacophony of prayers followed, with each member of the group making petitions of their own, explaining their needs. Finally, after an hour or so, the sun rose, and the icon of Santiago was held before the fire and the rising sun (Figure 3.3.). Rudy said later that the ceremony was carried to 'draw Santiago's spirituality' or *nawal* to the icon at the rising sun, since Santiago's *nawal* is considered to be 'hot' and powerful too.

Figure 3.3. At sunrise, the members of the group bless the local Santiago icon at the *Ventana Mundo* altar. © Alonso Zamora Corona

Afterwards, as in the plan laid at the preparatory ritual, at 7am Virgilio and Rudy went to the church, to bring the icon near the actual image of Santiago during the morning eucharist. Finally, at 11am, the local *feria* took place. The dancers began the performance with marimba music at the *convite*, the invitation ceremony to draw the attention of the people towards the arrival of the image. Santiago's image arrived at the main road at the *caserío*, where it received the respects of the rest of the members of the *cofradía*. The very same icon that Virgilio had blessed the prior night and in the morning with the visit to *Ventana Mundo* was then placed before the images of Santiago and Felipe, both on their palanquins, an act that is somewhat of a circulation of forces: by being blessed before the *wachib'al* or image of Santiago, the local Santiago icon was strengthened; however, it is also true that Santiago's image itself was in a way being strengthened by the contact with a duplication of itself, which has also received cult status, something we will touch on later. Finally, the image of Santiago followed on its way, while the local *convite* or amusement parade and dance continued, with merriment, food and music, which lasted well into the night.

Now, I have presented these rituals and celebrations in terms that are more or less 'detached', explaining why, how and when the celebration was performed. However, from an ethnographic point of view, what was particularly interesting to me was that moment of joking where Rudy passed from the usual humble and petitioning self that he would always show when engaging with ancestors and elders, to that of a certain assertiveness and power. As quoted, he said, 'Sooner or later there shall be authority over them' (*tarde o temprano, habrá autoridad sobre ellos*), referring to those reluctant to partake in the celebration. To me, the possibility that Rudy started making those assertions was the real outcome of the ritual, its manifestation as a construction of his place in the ritual and political life of the community. However, I do not wish to understand this assertion in the terms of a crude functionalism; that is, by saying that power itself is the actual *raison d'être* of the ritual, for it is not so, as Virgilio literally admitted: the celebration is just a factor, among many others, to be considered eligible for an appointment. Instead, I want to understand this assertion as something *contextually constructed*, framed by the ritual 'scaffolding' of sorts that precedes it.

From now on, I will concentrate on a detailed analysis of the temporal relationship of both rituals, through the lenses of the anthropology of time, the main question being the following: as anthropologists, we often talk about 'circular' or 'cyclical' time in regards to non-Western cosmologies, in a rather unsatisfactory way, as it has been pointed (Holbraad and Pedersen 2017: 17), so how do we overcome this state of our representation of time in non-Western cosmologies? What I suggest is that situations of apparent 'cyclicity' arise out of our misunderstanding of how time and causality operate in non-Western worlds.

The Anthropology of Time and Temporal Paradoxes in Maya Cosmologies

Now it is time to analyse both rituals and their relationship. The most important fact is that, through the preparatory ritual, Rudy was trying to influence the outcome of the second ritual. This is very clear: as I mentioned, during the ritual he started telling the spirits of the ancestors and gods to 'move' and compel the spirits of people and even inanimate objects, so that the second ritual would be successful. As to why he did this, a few ethnographic observations can help us understand further. As Rudy explained to me once, it is the ancestors (whom he considers as gods) who ensure that things actually occur in this world in the way they do; however, for the gods, all that will happen in our world has actually *already happened in theirs*. For example, Rudy told me that births, deaths and life events of people in the patrilineage are decreed by the ancestors and this is the reason why it is possible for him to know when and how they are going to happen. In the case of his own marriage, for example, he told me that he dreamed about it beforehand. He dreamed about going into a mountain, and finding a pair of ancestors sat over stones. He told me how the ancestors there performed the marriage ritual for him, thus making him already 'married' in the *Santo Mundo* before he was married in this world:

> The ancestors arrived in my dream, during the night they came to tell something to my soul, they came to show a wedding, my hands and feet were to be bound with a woman. I was going to get married. I dreamed how they (the ancestors) were sweeping the road, putting white flowers and singing. I arrived to a great female ancestor, she received me, gave me white clothes, gave me a bundle of flowers, she also said: your hands and feet are going to be bound, here it is your woman, here is your wedding, when you marry, this is your partner, she will be with you until you die.

Clarifying the dream in Spanish, he explained to me that the flower bundle in his dream was his own wife, which according to the ancestors would come from the direction of the calendrical mountain of the north, Pipil; he also told me that the ancestors told him in his dream *ya te casaste*, 'you are already married', referring to the future as something already past.[8]

Similar paradoxes are to be found in other examples from Maya ethnography. For example, Esther Hermitte, an Argentine ethnographer working with the Tzeltal of Pinola in Chiapas, Mexico, collected the story of Bricio, a man who was invited to participate in the local Indigenous government but was hesitant to do so, since he has illiterate. However, a shaman told him that *he already knew how to write*, not in this world, but in the world of the ancestors or *me'iltatil*.[9] These words encouraged him to learn to read and write, and eventually he became part of the local government, self-fulfilling the assertions of the shamans (cfr. Hermitte 1970: 70–73). Thus, both assertions, that of Rudy's ancestors and that of the Tzotzil shamans, are similar: 'you are already married' and 'you already know how to write'. Indeed, both are things that had already happened in the world of the ancestors, and then they became real in our world. This implies the idea that things that will happen in our future are already past in the perspective of the gods.

Thus, we find again the paradox of the 'strange loop': the ancestors set the future in motion because humans compelled them to do so in the first place, through the rituals. The world of the ancestors seems to contain the temporal world of humans in a sort of transitive relationship (our actions and temporal perceptions proceed from it), and in a relationship of hierarchy (they determine what happens here), but, at the same time, the actions of people can constrain or direct the actions of the world of gods. As strange as it sounds, a calendrical priest could say, regarding this situation: 'I made the gods do it' but also, the gods made the shaman do what he did (a self-recursive paradox in causality). Furthermore, it is possible to assert that the ritual tries to transform our future into the 'past' of the world of the gods, converting something mutable into something 'decided'. Thus, this is not really a question of 'cyclical time', but of inclusion of our time into another dimension, which is not completely temporal in our sense of the word. This leads to the crucial theoretical question: what is this 'time' where our time is contained? Perhaps here it will be useful to take a glance at how anthropology has dealt with the question of time.

The question of time has a rich and complex theoretical genealogy in social anthropology. The problem traces its origins to Durkheim's first sociological formulations (see Katovich 1987), but it made its appearance in social anthropology proper with Malinowski's remarks on time-reckoning in the Trobriands (1927), and then developed through classic examples such as Evans-Pritchard's fascinating insights on the relationship of ecology, social structure and time among the Nuer (1939, 1940), Leach's conception of 'alternating' time (1950, 1971), Lévi-Strauss' famous notion of rituals within 'cold societies' as an instrument for the abolition of time ([1962] 1966: 235–6; cfr. Gell 1992: 23–29) and Geertz's account of Balinese calendrics (1966), which introduced the question of a qualitative aspect of time, just to name the most relevant examples from classic anthropology.

The most well-known and relatively recent synthesis of the problem (and a classic in its own right) is of course Gell's *The Anthropology of Time* (1992), which concentrates on the notion of times series, which Gell took from the idealist J. M. E. McTaggart (1908, 1927). The time series are the two simplest forms in which the concept of time can be reasonably posited for analysis: the A-series, or time conceived as being formed by the contrast between 'past', 'present' and 'future', and the B-series, which emphasizes the contrast between 'before' and 'after' (Gell 1992: 151). Gell declares himself a supporter of 'the moderate version of the B-series position'; nonetheless, throughout his work he considers the validity of both series in anthropological analysis and contrasts their features and the different regimes created by them in economics, time-mapping, social structure and praxeology.[10]

In his anthropological theory of time, Gell only focused on the contrast between A-series and B-series time conceptions. None of these seem completely adequate to describe our conundrum. However, the original objective of J. M. E. McTaggart, the creator of the concept of time series, was to show the 'unreality' of time, or how both A- and B-series were inconsistent. For McTaggart, the B-series of 'before' and 'after' was free of contradiction, but it was not a true temporal series, because it does not imply change, only the ordering of events in 'earlier' and 'later' (cfr. McTaggart 1927: 12). The A-series, that of past, present and future, does imply change and, with it, temporality, but it does imply a contradiction, for past, present and future are 'incompatible determinations' (McTaggart 1927: 20) that must be predicated of every event within time at some point, and, furthermore, can only be predicated in function of each other (past being *after* present and future, present being *after* past but *before* future, future being *after* both), thus generating a 'vicious' regress (cfr. McTaggart 1927: 22), because we are using temporal categories to explain time itself. Thus, in McTaggart's view, both A-series and B-series are 'unreal'.[11] To surmount this difficulty, he proposed the idea of a C-series, a non-temporal order, and the illusion of time arising from our misperceptions, either the A-series or the B-series (McTaggart 1927: 30).

McTaggart changed his ideas on the nature of the C-Series during the development of his philosophy, but some basic aspects of it remained. One such point is that he considered the C-Series to be a non-temporal, unchangeable order of events:

> There is a series – a series of the permanent relations to one another of those realities which in time are events – and it is the combination of this series with the A determinations which gives time. But this other series – let us call it the C series – is not temporal, for it involves no change, but only an order. Events have an order. They are, let us say, in the order M, N, O, P. And they are therefore not in the order M, O, N, P, or O, N, M, P, or in any other possible order. But that they have this order no more implies that there is any change than the order of the letters of the alphabet, or of the Peers on the Parliament Roll, implies any change. And thus, those realities which appear to us as events might form such a series without being entitled to the name of events, since that name is only given to realities which are in a time series. It is only when change and time come in that the relations of this C series become relations of earlier and later. (McTaggart 1908: 462)

Thus, the C-Series is a sort of fixed series, like the letters of the alphabet. One popular view is that a C-series is 'an expression synonymous with "B-series" when the latter is shorn of its temporal connotations' (Shorter 1986: 226); that is, stripped from the idea of 'before' and 'after'. What does supply the notion of order in this series is that the C-Series is an 'inclusion series' (McDaniel 2020), in which terms are conceived as being 'included in' or 'inclusive of'. Finally, McTaggart speculated about a final point to the C-series, 'the term which is the whole of the series, the term which includes all the others, and is included in none of the others' (McTaggart 1927: 365). McTaggart believed that this 'whole' of time belonged to the future, the final truth of time, the stage where all temporal perceptions are contained in a 'truthful' way, which he called 'the futurity of the whole' (cfr. McTaggart 1927: 365). This is, in a way, a world without a future, only with past, a world of 'truth'. But let us leave McTaggart's ideas at this point.

My core argument is the following. I believe the 'ancestral' or 'cosmological' time presented in this chapter is, of course, not a 'circular time', nor anything that we can call time in the traditional sense (the A-Series and B-Series of Gell), but instead somewhat of a C-Series. This 'C-Series' world is perceived by us as determining our future but, from its point of view, we are in the 'past', or to be more precise, we are *included* in it: in that world one 'already knows how to write', the ritual is already done, or one is 'already married', all events that in our world belong to the future.[12] However, the crucial difference, in regards to McTaggart, is that in this non-Western case, the 'final order' is not only inclusive of but also somewhat included in our own temporality, in the form of sacred shrines that contain the access to the cosmological *Santo Mundo*, thus giving rise to all sorts of paradoxes, like being able to see the future (which is the past of the ancestors) or perform rituals such as the one presented in this chapter. Thus, this world, which in the Western philosophy of McTaggart is a final, and fixed 'truthful world', is always in a self-changing relationship with itself, somewhat like Escher's well-known 'Drawing Hands' lithography (1948).

Thus, going back to the ritual, what Rudy does in the preparatory rituals is to compel the ancestors to make sure the rituals happen in that temporal dimension in which our actions are already understood and decided. He describes the rituals before the eyes of the ancestor-gods, so the correct execution of the ritual is already anticipated, and then their appearance in our world under the desired guise becomes, quite literally, *a mere matter of time*. The result is a sort of equivalence between his agency and that of his *cofradía* with that of the ancestors and gods: who has obtained the money for the ritual? The *cofradía* and him. Who has paid for the music and the beer? The *cofradía* and him. Who has expanded and consolidated the social relationships that make the ritual possible? Them. But at the same time, for them *it was not them who did it*: it was the ancestors and the gods, *who had already done it*. The objective is that through the ritual they made sure that it was not them doing things, but the gods, and that things appeared as something 'already done' in the temporal dimension of the gods, which we could characterize as a sort of 'C-series world' in the temporal sense.

However, the point of the ritual is only not to perform the action, nor to make the gods do it, but to constrain the action in a certain way, so that its outcome is favourable and has a desired form. I would argue that this is like magic in Kiriwina as described by Malinowski, whose objective was not to create things, but to make them efficient and good (cfr. Malinowski 2002 [1922]: 125). However, instead of having a more spatial origin, as the Kiriwinan magic, which comes from the underground dimension or Tudava where the Baloma spirits live and which is brought back to our world by the agency of the enchanted breath of the magician (cfr. Mosko 2014), for the Maya this sort of magic comes from time, or more precisely from a 'strange loop of time': magic is the feeling that our actions are not flowing from our temporal own level, but as something that unfolds from a dimension where time is perceived as past, and that we perceive as being future.

In conclusion, here we not have only a reversal of causality in regards to space, as it happened with the example of the destruction of the sacred altars, *but a sort of reversion of how causality flows in regards to time.* In our world, time regularly flows from the past into the future, thus making the past fixed and the future variable; however, the objective of the ritual is to make of *our* future something *fixed* and inevitable, and the past of the gods, the true past, which is our future, something that we can influence. Thus, the true aim of the ritual is, so to speak, to make time in our world flow *from the future*, rather than from the past. Usually, causation in the K'iche' universe goes from the ancestors to men, but here men compel the ancestors to make themselves act. The result is a loop in time. The order of the ancestors is non-temporal in relation to ours, because *there is no change in it*; it is already decided. It exists in a way that we cannot understand as temporal: it has no real 'future' in it, only 'past'. It is a C-Series, in the vein of McTaggart: a 'whole' that contains what we call time as a 'truthful' non-temporal order, instead of an order of mutable perceptions. However, since both this non-temporal world and our time are in a relationship of hierarchical entanglement, it is sometimes possible to know this world, and furthermore to try to place our future events as something belonging to this order, in the precise way we want it. Thus, the aim of the ritual is also to bring time into the world of 'eternity', rather than only bring 'eternity' into the order of time. Hence arises here the 'illusion' of cyclical time, which is not cyclical at all; it is a 'strange loop', an *entanglement of time* between our world and the world of the ancestral gods.

Notes

1. A somewhat similar epithet has already been recorded by Garrett Cook, who mentions that Santiago was invested as 'captain of the whole world' by the old *principales* (2000: 65). Actually, this epithet refers not to our world (or at least not only), but to the *Santo Mundo* or 'Holy Earth' itself. Although eminently linked to the earth and the underworld realm of the dead, the *Santo Mundo* comprises also the sky, as evidenced by the prayers recorded by Schultze-Jena, where ancestors (who are *mundos* or 'earth gods' themselves) are called *rawajal ri kaj, juyub'* 'lords of the sky and the mountains' (1933: 19). The German ethnologist also recorded prayers where the god of the earth is said to be able to communicate with the Moon and the stars (cfr. 1933: 53).
2. Cook had already stressed the importance of the changing of Santiago's robes in his cult (2000: 88–89). I never got the chance to corroborate Virgilio's and Rudy's assertion, but it is true that the colour of Santiago's clothes during his parade changed from one year to the other while I was at the town, one year donning a black suit, the other a blue one. Rudy's testimony regarding this particular is this: 'When he dons black clothes, it will rain all the year through. When it is blue, rain won't fall. Our fortune for the year can be seen there. The *alcalde* of the image, being very devout, is told by the image itself which robes it wants to wear, because Santiago has many changes of clothes, not only one, he has many, so he comes to warn him, to tell him the clothes he must don, and they (the members of the *cofradía*) go looking for them. His clothes are changed at midnight.'
3. For more on the concept of *wachib'al*, see Zamora Corona 2019: 112–15.
4. 'I went there, I was invited to be a part of the ceremony. We slept at the chapel, and the *alcalde* of Santiago or the image's *alcalde* wouldn't allow us to sleep before the image, because his horse comes out; it can trample you and you can die during the vigil. One can sleep there, but at the side. I have heard it. One night, I had been keeping guard on him and I saw him rising in my dream, giving water to his horse and coming out, I listened him opening the door, coming back and returning.'
5. The Guatemalan *convite* has been defined as a comic parade (*desfile bufo*) where participants wear masks and disguises and often dance in order to invite neighbours of a community to participate in local festivities, hence the name *convite*, which means 'invitation'. They have been celebrated since colonial times to announce festivities during the itinerary of religious processions, and now are characterized by the incorporation of characters from pop culture and movies (García Escobar, 2007; Taube 2013: 102; Cook and Offit 2008: 96–118).
6. The number is of cosmological significance to Maya cultures, often meaning perpetuity or being innumerable (Stuart 2012: 125).
7. The K'iche' language is full of loanwords from Spanish; some of them are archaisms. However, I have found no clear precedent or explanation of the word *vasalla*, which is used to refer to a company of dancers who have taken a vote to perform before the saints; it can refer to Santiago's dancers, like here, or to the performers at traditional dances like the Monkey Dance (Cook and Offit 2013; Cook and Offit 2008: 55–95) and the *gracejos* (Cook 2000: 171–82). The word sounds like *vasallo*, 'vassal', but the singular feminine marker '*a*' does not seem to fit here.
8. This rhetoric device is reminiscent of the so-called 'prophetic perfect tense' in Hebrew biblical literature, where a future event is spoken about as if it had already happened (cfr. Klein 1990).

9. 'Shaman: Bricio, you want us to beg you to participate (in the government), but you evade our petitions. But we know that in the other world you are very intelligent ...
   Bricio: I don't know how I am in the other world ...
   S: You don't know how to write?
   B: I don't.
   S: But in the other world, you know. I have seen it in my spirit' (Hermitte 1970: 70).
10. It is also important to mention that, after Gell, Marilyn Strathern has proposed the idea of 'fractal time' along the lines of 'Cantor's dust' (cfr. [1991] 2004: 3), a notion that has been taken and further developed in recent times (cfr. Dobroski 2019).
11. McTaggart's argument has been recently construed as an argument about tenses, not against time, in the vein of analytical philosophy (Mellor 1998), while other recent critics have argued that an A-series is not actually contradictory (Hansson Wahlberg 2013; Ingthorsson 2016); however, the truth is that the debate between proposers of the A view and the B view continues unabated and unsolved (cfr. Ingthorsson 2016: 1; Smith 1994).
12. Of course, in the ritual Rudy speaks in a temporal manner: 'we hope that you come with us', 'we hope that this will happen' and so on, but as the examples of the dreams show, what he hopes is not that the ancestors and gods will actually come down from Heaven during the ritual, but that the ritual is already 'performed' by the time it actually happens in our world, making it inevitable and making it happen in the precise way he wants it to happen.

4

'If You See Me in Dreams, I Can Harm You'

On Witchcraft

The present chapter represents a break with my former ethnographic narrative, although not with the general concern of this work with cosmology and with its social consequences. It will leave the more localized narrative of the Ramos family, and instead depict a topic that is of concern to all people in Momostenango, at least in the words of my many interlocutors: sorcery. Thus, instead of depicting a concrete ethnographic voice or interlocutor, this chapter will opt to present somewhat of a polyphony of voices. However, as the reader will see, this blurring or lack of focus has a certain advantage: it depicts, I think, how sorcery and 'witchcraft' appear in my fieldwork experience, as more of a recurrent but acute preoccupation than a topic that I engaged with in a continuous manner.

As a topic, witchcraft has been said to be 'at the epistemological centre' of anthropology (cfr. Kapferer 2002: 1), and yet seems to mainly consist of 'unanswered questions' and 'indeterminacy' (Niehaus 2012: 717). Despite the continued presence of functionalist approaches in the study of the subject, which mainly consider witchcraft as an expression of personal and social anxieties (cfr. Gersham 2022), in recent years the topic has opened towards approaches that take more seriously the reality of those practices in the lives of informants and their epistemologies about it: thus, as Harry West notes, to the dismay of anthropologists, informants adamantly dismiss the anthropological ideas on sorcery and insist in its reality (2005).[1] The topic is made even more complex by the fact that witchcraft often stands in a problematic relationship, not only with anthropological interpretations of it but also with the native cosmologies that could frame it. As Bruce Kapferer aptly remarked, witchcraft can be considered, cosmology-wise, as a place of invention, and it often proceeds by 'breaking apart elements of other cosmological schemes' (2002: 21). Thus, in a theoretically challenging way, witchcraft is not only tremendously conflictive within societies and within anthropology, but even within native cosmologies themselves.

Of course, it is not my intention to offer any definitive answers to questions as old as anthropology here; it would also be absolutely misleading to ignore or minimize the numerous treatments of the subject that precede this book. Even the proverbial pages by Evans-Pritchard regarding Azande people unwittingly setting fire to the thatch of their huts, or having their granaries suddenly collapsing, and then 'explaining' those particular happenings away through witchcraft ([1937] 1976: 22), pages that are the departure point for all subsequent anthropological studies of witchcraft, are charged with powerful interpretive possibilities undetected by our usual functionalist readings, as Martin Holbraad has recently suggested (2020). Similarly, influential subsequent accounts, such as Favret-Saada's masterful works on counter sorcery (1980), West's notion on 'animated metaphors' (2007) and Kapferer's idea on autonomous logics of sorcery that break apart with everyday reason (2002), are of a critical and enduring value and will of course inform my own engagement with the topic. Furthermore, unlike the aforementioned authors, my engagements with witchcraft were fragmentary and unsystematic, so it is impossible that the current understanding of the topic can be truly deepened by my limited engagements with it. However, despite this setback, in this chapter I will try to propose a somewhat experimental approach to the topic that is born out of my particular ethnographic experience as well as the general proposal of this book.

This chapter is, as mentioned, somewhat experimental in its approach. Instead of introducing the topic as some 'irrational' explanation or belonging to 'the other', I will first introduce a very brief ethnographic encounter with it, one that was personal, although it cannot be considered as a real case of witchcraft; instead, it is an engaged analysis of the statement of an informant regarding witchcraft, a statement that precisely seems to embody what Bruce Kapferer has characterized as the raw power of 'standing outside reason' inherent in beliefs and practices pertaining to the occult (cfr. 2002: 11). This statement will be the starting point for different engagements with witchcraft. What I will try to argue is that different statements regarding witchcraft can be seen, conceptually, as describing a change in the perceived flow of causality, so instead of perceiving actions, and misfortune in particular, as flowing from the expected order of things, we see them as flowing from the influence of a malicious agent, the sorcerer, not in a regular fashion, but in a cosmological one. The victim, or the society around it, perceives its doom as being caused by a person, rather than as being brought about by the random whims of the cosmos. Thus, witchcraft will be defined here as a sort of 'usurpation' of causality by the sorcerer, but it is curious how this change in the quality of the causal sequence sometimes arises within the victim itself. Also, regarding the idea of the cosmological locus of witchcraft, we will see that the Maya insist that sorcery expresses a certain part of the cosmos and its 'nature', although it is precisely its margins, the powers that aspire to destroy the cosmological order. Thus, cosmology as a place and an inversion of causality play a rather important role on sorcery too, which this chapter will seek to explain.

'The Night Shall Vanquish the Day': A *Prolegomena* to Sorcery

I am alone on a road, on the way to meet an informant, just at the beginning of my ethnographic fieldwork. Being immersed in isolated and extended fieldwork for the first time, I wondered what was going to happen. Would I commit some kind of mistake? Would he turn me away? While thinking about this, I arrived at a crossroads on a mud road in the mountain to meet Alberto, who was leaning on a derelict adobe wall. He was going to show me around the shrines of the town. We greeted each other and started talking about life. At the time, his wife was pregnant. In fact, Alberto told me that she was expected to give birth in a few days. The doctors had told him that most probably it was a girl. But he told me he was convinced it was a boy. He explained to me that he had seen the *nawal* of his son in dreams, as a man, for a *nawal* has not only an animal form, but sometimes a human one too. He told me that he was convinced that his son would be strong, that he would have a strong *nawal*, just like him. He then told me that the ancestors had already told him that his son was going to be his successor: '*Dios Mundo* has decreed that he is a man.' And then he added more about why he knew this:

> In dreams I have seen the *nawal* of my son, he is a man. I know it because of the skin of his *nawal*, the *nawal* of a man is dark-skinned, the *nawal* of a woman is often clear-skinned. The strength of the *nawal* in relationship with its skin colour is because of the difference between the night and the day. The day is like the sun above, it is weaker, it is clear, he is a *gringo* like you,

he cannot stand being seen, that's why he dazzles us with his rays, because he doesn't like being seen. The night, the moon, however, is patient, she is an Indigenous woman, she bears the burden of being seen, she fascinates [*fascina*]. The night is active and patient, strong. The sun is like the *canche* [the white man], who is weak, passive: he sees. The dark skin is stronger, for it is like the night, and the night shall vanquish the day in the end. This is the reason why, if you see me in dreams, I can harm you. If I see you in my dreams, you can do nothing to me [*Por eso es que si tú me ves en sueños, yo puedo hacerte daño. Si yo te veo en sueños, tú no me puedes hacer nada a mí*].

While uttered in the context of a non-hostile exchange regarding life, these words were really disturbing for me. Frankly, to me they were almost like an arabesque: one can get lost in their meaning, just dazzled by recalling them once, and then again. Besides the literal allusion to the idea of dazzlement in Alberto's words, anthropologically this ethnographic moment still exerts on me a dazzle-effect similar to that described by Marilyn Strathern, for it certainly was linked to a sort of 'revelatory sense of occult practices'.[2] These words were a literal subversion of the anthropological gaze and of the role of the anthropologists, that 'participant observer' who does not really participate that much; like the all-seeing eye of the sun, and yet weak precisely because their job is to see and not to be seen (an easy, 'weak' job). Identifying himself instead with the moon, which 'bears the burden' of being seen, Alberto was in a way reverting the power of my gaze over him into a deeper, more powerful, hidden power, a power of 'being seen' while not really being touched nor affected. A more powerful gaze, one that is not that of the anthropological engagement, was alluded, precisely that which is alluded to when informants discard the futility of anthropological explanations and talk about the reality of witchcraft. After these enigmatic words, Alberto and I started talking about many other topics, and at some point (when asked by me) he started talking about sorcerers (*ajitz*) and sorcery (*itzinel*) in detail:

There are sorcerers, those who perform evil. They know the calendar. They become animals, their soul exits their bodies and it enters in an animal. People hear steps in the roof, animals screeching, howling, singing and crying outside. The heart of the witch exits his body and he send forth the *win* to do harm,[3] he uses the days of the calendar to pray to the lost souls (*las almas perdidas*), those who died in violent deaths, those who disappeared, those who float in darkness and in night. 'Come with me, come,' they say to the lost dead, they use their spirits to come out in the night. Hanged men and suicide victims are gods of the sorcerers (*los colgados, los suicidas son los dioses de los brujos*). They invoke them, they use those souls, the lost souls. In the rooms where they make their offerings, they have their instruments,

> in rooms without windows, in complete darkness, where light won't enter. They ask for permission with their ritual implements in darkness. Sorcerers are loaners of spirits, those who lend their spirit to another body are witches.

Of course, Alberto was not a sorcerer or *ajitz* in any sense, but in a way his first words, about the sun and the moon, and the game of gazes between the anthropologist and his 'informant', could be said to convey something about the ability to do harm beyond natural means and, furthermore, to the relationship between the sorcerer and the victim too. The victim could perhaps see their bewitcher in dreams, but will be impotent to harm them, as if the sorcerer was untouchable or made of thin air. The sorcerer, meanwhile, can see and act without being harmed. If one thinks of the allusion to the sun and the moon, the difference between the 'sun' of the anthropologists and the 'moon' of the informant is this: the anthropologist can see, but is weak, he cannot stand 'being seen' in a different way, and tries to dazzle the look of his interlocutor away; meanwhile, the fascinating 'other' is seen, and it is strong and can stand being seen, but the difference is that he dazzles in another, deeper way: not by make us avert our gaze, but by means of *fascinating* our gaze.

Finally, what these words allude to is an asymmetry in causality, a certain hierarchy. The sorcerer can act without being harmed; the victim, on the other hand, cannot causally influence the sorcerer. As we will see later, this can only be achieved through counter-sorcery. Thus, in a way, the initial situation of sorcery is like that: a nefarious influence that acts without being acted upon, an unequal relationship in terms of causality. For now, let us leave this point here, and proceed with Alberto's testimony on witchcraft, conceived in a cosmological manner.

The K'iche' Cosmology of Witchcraft

Once, many months after our initial encounter and when we were already friends, Alberto and I were talking again about witchcraft. Instead of just talking about cases of bewitching or rituals, he gave me a long explanation regarding what can be portrayed as a cosmology of witchcraft. Alberto told me that his great-grandmother lived near Quetzaltenango, to later arrive at Momostenango. When he was a child, she told him the story of Juan No'j, a very peculiar character of Guatemalan folklore, on which a few words are needed before jumping in to Alberto's testimony. According to local folklore, Juan No'j was a sorcerer or *ajitz* who achieved great wealth and power in life through his dealings with the underworld, but eventually died and, as an afterlife punishment, became the underworld master of the Santa María volcano. It is said that he lives inside it, and has numerous souls that are enslaved to him, people that sold to him through sorcery during his life. He has a beautiful hacienda, but he must always rebuild it because of the constant eruptions and fires inside. It seems that Juan No'j is connected somehow with the cult status given to the neighbouring Cerro Quemado ('burned mountain' *k'atinak juyub'*) volcano, now extinguished. No'j[4] is the *nawal* or calendrical essence of the Cerro Quemado. The Cerro Quemado volcano contained an ancient place of worship that was moved to the Santa María volcano when part of the Cerro Quemado was destroyed by an eruption in 1818. There is a cave at the foot of the Cerro Quemado volcano with many altars to Juan No'j, where people carry out ceremonies to ask him for money or health, but also to make witchcraft workings (Méndez Bauer 2015: 78–80). According to Benson Saler, in the K'iche' community of El Palmar, Juan No'j was considered a powerful *dios mundo* or earth god and could possess *nawal mesa* ritual specialists in order to give answers to people (cfr. 1960: 143).[5] A similar cult has been documented by Juan Porres (1992) in Samayac, the so-called 'capital of witchcraft in Guatemala'.

Since Alberto's grandmother was from Xela, she knew very well the lore about Juan No'j and his abode in Cerro Quemado, and told him about it in detail:

> My grandmother said that when they arrived from the coast, at 5am, they crossed on mornings at the Santa María Volcano in Xela. My grandmother said that she cried then, because of fear. 'I couldn't sleep, because of the fear, the people at the volcano crying for help, they were tired, some said, "We are burning," others were saying, "How heavy are these stones!" others said, "We are tired of holding the volcano," others laughed, the cries of babies, of old people were heard … and the sounds of the animals, the animals of the mountain, the snakes, they were talking: get us out. If any animal calls you like that, don't bring it out, if you do, the animal will become a stone, it will take that form immediately …
>
> The people that cry in the volcano are sold to evil, they got money from the earth, they ask, and the volcano gives money, it gives everything, sometimes the volcano would say, "No, I don't want old people anymore [easy victims of witchcraft], I want one of your sons," and people give to him [the volcano god] what he desires, they bring new-born babies and sacrifice them, so their souls are there too. All those people end up shovelling the fire, the stones, that's why the volcano won't ever fall. The people that are inside are holding it, the volcano, be sure of that, it will never end. My

grandmother said: each roar of the volcano is the arrival of a new soul in it. That's the story that my grandmother told.

Alberto further explained that, just like there is God and his opposite, the devil, the earth as a benefactor or protector has its 'contrary' or 'enemy' in the figure of Juan No'j and similarly aggressive *dioses mundo*. The ancestors or *nantat* are those who sustain and feed people, and similarly the earth lords or *rajawal ri uwachulew* are invested in sustaining a cosmological order. However, the Mundo has also chaotic forces, personified by the *dueños* or *dioses mundo* of the volcanoes (*xkanul*) and the sea (*palo'*). Sorcerers are servants of the powerful spirits of these places, and make offerings at special altars in the vicinity of volcanoes and the of sea, for both the sea and the volcanoes want people to 'carry waves and fire' as slaves. The sorcerer makes offerings (sometimes bloody) to them, and they grant extraordinary gifts to their servants. 'There are altars', an *ajq'iij* told me once, 'where you cannot enter without offering a good morsel [*bocado de carne*] for the underworld.' At first, the earth asks for small things: rabbits, quails and turkeys. But then, sacrifices become more expensive: goats, sheep, bulls, and even people:

> Those who kill animals are sold to the volcano lord, they do that in order not to go there themselves, because the volcano says to them: come with me, and they reply, I don't want to, but I will always feed you. The volcano may say: I want the flesh of a pigeon, it has to be black or white, now I want a sheep, a black one, and they have to search for it, pay its price. They may get a bull, here is the bull, in order not to go. They are already sold, but not completely, because they feed [the volcano]. They feed the earth for evil, to get things for themselves, that's why they put before them the lives of those little animals or people, not their own. Yes, they burn at the altar of the two-handed cross, the Mesa del Santo Mundo [an altar in Momostenango, used for evil]. They burn there, but not everyone, just the ones that sold themselves. One can speak directly with the earth there ...
>
> If you go the sea, there are altars near the sea, if you go there [to burn an offering], the sea will give you something, but you will stay there to shovel the waves when there are sea earthquakes, because there are lots of people chained at the sea. They are already sold. That's the story of how one wants to become someone overnight. The god of the earth, his counterpart is evil. They compete, so to speak: the volcano, the sea, with the earth god, to win more people. They give their lives.
>
> This earth, it is in peace from some years, but only until you have to sacrifice something there, although not necessarily people. Sometimes it is pigeons, sheep, even a bull, in the enchanted places. They are those who are sold to evil. Because the earth gives money. They go to the volcano at Xela or any other, and say: we bring you a person, or a bull, we want money, we

> give this in payment. The *Mundo* says 'It is fine, but when you die, you will come here with me.' It says, 'The day you die you come to shovel fire and stone.' They say, 'We will come to help.' They do their ceremonies and many things appear to them.

Although Saler denied the existence of a pact between the sorcerer and the underworld powers among the K'iche' (1964: 320), the prevailing opinion in Momostenango suggests otherwise. The most powerful *ajitz* are considered as those who make a pact with the spirit of a volcano or the sea in order to fulfil their wishes. The sorcerer becomes more powerful and rich with each sacrifice. Eventually, the sorcerer becomes untouchable, since 'only the earth itself can kill him, no bullet or machete can kill him', a boast that some powerful *chuchqajaw* make too. Weak sorcerers fail and becomes indentured slaves of the underworld powers (workers in the hacienda of Juan No'j, moving shovels of fire or waves), but the strong ones can rise to become *awajmundo* themselves, guardians of the underworld, *nawales* themselves. Alberto recalled to me once the story of a sorcerer who made sacrifices of this kind and eventually was granted lots of money by the Mundo:

> There was a man living near here, further down the mountain, he did his ceremony there, and one day a little animal appeared, he followed it and a clay pot appeared. When he arrived to his home with it, he placed his altar in a corner, he performed a ceremony, covered it with a white sheet, and at midnight he heard someone pushing the door, at sunrise, when he saw the pot, it was full of money. He took the money, bought lots of things, built homes, how many things he had, even workers to bring him food. But when he died, at midnight, his family opened his coffin, and they found nothing but animal bones, only deer bones; he had become a deer, he decided to become a guardian of the Mundo, he went to take care of the volcano, lots of people want to hunt animals, but they can't, because those animals are souls of people, they disappear, they take care of the place.

Animals like the *venado del mundo* (deer of the underworld) are the spirits of powerful sorcerers who serve the volcano or the mountain as *awajmundo* themselves; they are masters of game, and they are usually portrayed as punishing hunters and trespassers to sacred places (*encanto mundo*). Also, the powerful *dueños* at some mountains are the ones who grant sorcerers the gift of using spirits or co-essences in offensive sorcery, which are called *win* (cfr. Zamora Corona 2019: 125). Alberto then told me of a neighbour of his who was a witch and had a *win* spirit. This woman was a *q'eqa mesa* or 'black table' ritual specialist. The *q'eqa mesa* feed the altars where the 'forgotten ancestors' live, for they are the most powerful, being the most hidden:

> One must go at midnight, one must offer good food; for example, a black bull without blemish, a black rooster, the *dioses mundo* are the hungriest, their sacrifices take long time, they sometimes take place at volcanoes. Their days are 2 Aq'ab'al (2 Darkness), 1 Iq' (1 Wind). It is the hunger of the underworld. They deal directly with the mountain, unlike the common *chuchqajaw*, therefore their work is strongest, they deal directly with *El Mundo*. That's why their work is more demanding, they ask for more, they are harsh.

Thus, the duty of these specialists is to feed 'the old ancestors, those who are never mentioned, those who hunger, the forgotten'. The witch mentioned by Alberto was a black altar priest. When she received the *vara* or divining paraphernalia, she was told by the priests that she could make offerings at any place, without fear, at any time, even at midnight. He recalls her boldness by making midnight offerings at remote altars where others would not dare to go. He told me that, after her initiation, she invited him to see his altar. On her altar were Juan No'j and a black-clad idol in chains.[6]

The cosmology depicted by Alberto can be surely considered akin to what Kapferer has described as being part of the irreducible, disjointed and conflictive character of cosmologies associated to magic and sorcery (2002: 11). This cosmology of powerful and dangerous beings that are owners of the fringe spaces of the cosmological order (volcanoes, seas) is not only characterized by lurid tales of sorcery, but is also conflictive within the very cosmic order it stands on. It represents a promise of an unstable and dangerous power that is self-undermining, conflictive and uncertain. While this depiction is certainly representational, it is nonetheless important to understand how sorcery is not merely a set of practices, but it is embedded within a cosmology of its own, which stands apart from other cosmological schemes. Now, I would like to follow up with a testimony from an actual sorcerer, or rather, a 'counter-sorcerer', in order to present the perspective of an actual practitioner of so-called 'offensive magic' in my account.

A Counter-Sorcerer Speaks

As mentioned before, Alberto was not a sorcerer, despite his extensive knowledge of the topic, which came from both his acquaintance with the lore of Xela and his own experience with other ritual specialists. But I knew a sorcerer, at least in the sense of a person who could harm other people with ritual means, although, in the strict sense, he was more of a counter-sorcerer or 'unwitcher', in the vein described by Favret-Saada's masterful work *Deadly Words* (1980). My encounter with him was brief, and he just granted me a single ethnographic interview. But the story is worth telling, especially to present my reader with an actual portrait of a man involved with the 'circuit' of sorcery.

The encounter occurred some months after I started my fieldwork. After a night of conversation and drinking with the foreigners at the clinic in the faraway rural hamlet of Chwinajtajuyub', I was late arriving at Chunimatux village, where I promised to impart some English lessons to children at 9am. After taking a *flete* or old pick-up to the town centre, I decided to take a tuk-tuk, the sort of three-wheeler auto rickshaw that is very common in rural Guatemala because of its versatility and low price. Surprisingly, the driver was not a kid or a young man, the most common operators of this kind of vehicle, but an old man with a voice that was both sardonic and frail. As frequently happens on tuk-tuks, he asked me about my origins and what I was looking for in the village. I explained to the best of my ability that I was trying to understand calendrics, altars and traditions. He then told me: '*Yo quemo*' ('I burn'), the usual expression to indicate that one is a calendric specialist of some sort. He explained that he had a large *ekomal*, his 'burden', or the number of altars he sustains or feeds: 'I burn in thirty different places' he said, while, making the gesture of placing a tumpline over his forehead, an instrument so commonly used even today to bring merchandise from the villages to the market, but which also, in Momostenango, denotes the telluric deities, the Mam, who themselves bear the mountains and the days of the calendar on their back. He invited me to his home to have a chat about his work as *ajq'iij*, and we agreed to see each other the next day. The name of the man was Genaro.

Genaro is a merchant, with little agricultural land. He lives near the paved road in Barrio Santa Ana, which eventually leads to Pologuá. He does not really live in Momostenango much of the time; he travels to the capital and to the coast, always doing business. He has not passed on his divining gift to his children, who do activities unrelated to their father's otherworldly craft. Genaro is an *ajq'iij*, but he does do the same kind of work that others do. He is regarded by many to be a sorcerer. The day after our first meeting, I entered Genaro's home and was greeted by his wife on their patio. He immediately led me to his personal altar, such as many other K'iche' *ajq'iijab'* have in their homes; however, the room was much larger than that of the Ramos Ajtún family. An oversized figure of Maximon, seated in his chair, dominated the place with its presence, and many duplicate images of the same saint, a Virgen de Guadalupe, the Patron Santiago, San Felipe and oversized holy books, as well as black magic European grimoires, lay at his altar. Genaro explained me that over thirty years he has been 'playing with the *tz'ite*', by which he meant being a diviner or *aj'qij*. His true specialty, though, is counter-witchcraft. He is extremely busy; from time to time, he is interrupted by the calls of clients, which he answers using an old Nokia 3310 phone. For example, a man called him as we started the conversation; he had just crashed in Guatemala City and wanted to know 'who did the working on him' to have his revenge. Genaro wrote the man's name in an old appointment book, crammed with scribblings and photos of his clients, making sure to perform the necessary divination for the issue in time.

Shortly after this, Genaro started talking about his life. He was born in Momostenango, but eventually left for the capital, where he became a merchant; however, the high crime rate and extorsion there became a huge burden to his business. His fondness for drinking only made things worse. Eventually, in his own words, he 'lost everything'. Desperate, he consulted with the *ajq'iijab'*, who told him his sufferings were the result of an unanswered call from the Mundo, since he had the gift of the blood, a gift 'not for himself, but for others'. He explained that the 'speaking of the blood' guides his acts; in fact, the reason he agreed to talk to me was that 'the blood told him I had a true interest' in his craft. However, he was not initiated within his own patrilineage or at a group of *ajq'iijab'* related to it; he had to go to the coast, the proverbial place of sorcery, Samayac being the 'capital' of witchcraft in Guatemala. I have already elaborated on why the K'iche' strongly associate sorcery with both volcanic and coastal regions; it will suffice to say that, in a way, Genaro resembles his own patron saint, Maximon, for he is a merchant too, and he 'buys and sells souls'.

> It is the work of the Mundo. His work (is) money, money. And he kills too. If you don't feed him, he kills, if there are no candles, he kills. He wants his spiritual food. He gives money, but also misfortunes. This saint ... is a buyer and a seller. You know, who sold the lord? Who did it? Simón Judas. He received some money, the silver of the Jews. But he regretted it. He took the money, but he said: 'I don't want it anymore.' 'A sale is a sale,' said the Jews. He went to kill God, he went to kill Jesus. He was a buyer and seller, oh ancestors. This saint ... people say he is wicked, but he is not. He is a mirror, an example. God abandoned him. He does miracles, because God forsook him.

Maximon here is presented essentially a merchant, just like Genaro. His power is conceived in a catholic way, as a *speculum*, an *exemplum* of his fallen condition; he is powerful, he can sell souls, because he sold God himself. However, he also bears some of the characteristics of other K'iche' deities: he hungers for offerings, and he brings death to those who do not feed him. Two days belong to him above all: Kame, the day of death, and Tz'ikin, the day of wealth.

However, Genaro acknowledges that Maximon, as a 'Catholic' saint, is 'recent'. He says that the 'religions' in Momostenango are like *tz'ite'* seeds, as he takes some seeds from his divining bundle, placing them on the table. Protestants (*ri evangelio*) and Mormons are like fresh or 'green' seeds (*raxa tz'ite'*), immature, just recently arrived (corresponding to the altar hierarchy mentioned above). Catholicism is like the red seeds (*kaqa tz'ite'*), older, more powerful, stronger. Finally, the '*quemadero* religion', the altars, is 'the oldest, the most powerful'; it is like the black beans (*q'eqa tz'ite'*), as if they were charred by fire. 'They are the ancient things, the things that are like stone.' He began talking about the use of the calendar in order to attack, hurt and kill. He brought out a calendar and started making me read the names of the days aloud. First, Ajpu, symbolic of the dead. He said: 'I burn in the cemetery, where the dead belong. The *komon*.' In Momostenango there is a native altar in the cemetery, not far away from where the oldest church in town also stands, the *Iglesia del Calvario*. Appropriately, the altar is often called *kalwar*, but is also simply called *komon* (from the Spanish *común*, the common grave).

At that place lives *ajaw kalwar*, the lord of the dead, also the patron of the day Ajpu, the twentieth day of the Maya calendar, as we have seen in the chapter about calendrics. Kame is death itself, but Ajpu is the lord of the dead, and is considered to signify the divinized ancestors in divination. 'I go to the cemetery to call the dead before my enemy, so he sees the dead in the night, so he is frightened' and, thus, ends up joining them. Another day for witchcraft is Kan, which is used to ask for accidents, inflicting wounds, giving bad luck, like the one that the man who crashed in Guatemala suffered. Kame, associated with death, is to ask for killing, 'that his dog, his business, his animals or him may die':

> That day is a *nawal*, it is strong, a snake can arise from it, to bite. Let's say a tuk hits me. I then go to the cemetery, and call Ajaw Kan. Oh, dead one, know that a person hit me in the street, I know him, know where he is, but Mundo, you know where you son is.
>
> Lord Kan, take revenge. Lord one Kan, two Kan, three Kan … thirteen days. That the snake may rise tomorrow. He hit me, he robbed me, Lord Kan, arise. I don't burn only one day, but thirteen or nine days, and the snake will appear to him in the street. I summon the snake, a spiritual one, and it goes through the air [he gestures the snake rising from his magic book]. Let's say he is lying on his bed. He sees the snake in the night. He is frightened: the illness comes, cold and hot come to him. He steps on his grave [laughs].

The appearance of animals as bringers of misfortune and death (*lab'aj*, 'omens') is not uncommon in K'iche' witchcraft, and sometimes is considered as an attack of the *win* spirits or co-essences of a sorcerer. In this case, however, this sorcery is not conceived as a transformation of sorts (as in other stories, where the sorcerer actually becomes the *win*), but more as the sending of the underworld spirits to terrorize or even kill a rival. These *win* spirits are identified with the souls of the dead in Genaro's conception. According to him, nine or thirteen days of burning offerings in the day Iq' are needed to call the *xoch*, the owl of the underworld, a powerful *win*:

> If there is a bad person, we do nine, thirteen days of workings, then we make the animal rise against him. The animal will arrive, the owl, to call for him at night, at one o'clock in the morning [Genaro then hoots like a screecher owl]. It arrives [saying] may this son of a bitch die! He then departs for the graveyard, the call comes from there, you must raise the dead, for those animals are the souls of the dead.

We proceed to review other days of the calendar together, although not all of them have uses in sorcery. The day Kiej ('deer') is used to call the 'deer of the underworld' against an enemy. Garrett Cook (2000: 115) gathered some testimonies regarding this creature, which is also the *nawal* or soul of some altars and a protector of game; it is a huge and unnatural deer, sometimes with four eyes. For many of my informants, its appearance spells certain doom:

> 'Deer of the underworld, it is terrible, it is a son of a bitch. A son of a bitch. If you find that animal, you are dead, before God and the earth you are dead, if you cross him, it's not good luck, if you receive him, you will go to the graveyard, to rest.'[7]

Finally, Genaro finished his exposition on sorcery with some information regarding the role of the 'speaking of the blood' in it. He told me that in order to find the victim, he had to divine his location through his blood. As I have explained elsewhere, the blood of the *ajq'iij* is somewhat of a microcosm for it contains the things that are at the Mundo. The location of a person can be divined in that way too, although Genaro was not willing to go into more detail. He said that, in order to send the dead or the animals of the underworld to his enemies, he had to 'see' where they are. In a way, it is as if the underworld powers travel guided by his own blood. This fact fuels some geographic interest in him. During our conversation, he started asking me about faraway countries where his clients lived, like the United States and Spain, showing me a map and urging me to pinpoint the location of towns. He took out a photo decked in the pages of an old, Western-occultist grimoire, and showed me the image of a young Guatemalan lady, clad in her *huipil* and traditional K'iche' skirt before the Plaza de Cibeles, in Madrid. He explained that she is a client of his. He then asked me to pinpoint Spain on the map included in his old appointment book. He looked pleased, as if now he knew where to find his next victim, perhaps an enemy of this lady.

Finally, Genaro received another call, an urgent one, although I could not make out the subject of it at the time. He had to leave at once; we exited the altar room, and we parted for that day. After such a revealing exchange, Genaro became wary. He was still polite to me, and often I would come to find him, sometimes buying food in the market with his grandchildren, sometimes burning offerings at different altars in the town, especially at the Puja'l Santiago Altar. Even one of his sons, all grown up, is a tuk tuk driver in Momostenango, and I spoke to him once or twice (not about sorcery, of course). But Genaro would not ever talk again about his ritual business. 'No, I'm too busy', he would usually say, differing when I suggested potential interviews through a whole year, until I gave up. Perhaps persuaded by his blood, which guided him in so many different affairs, Genaro decided to hide his craft from unwanted eyes once more.

With this last testimony, we can start working out the interesting exchange of perspectives and agencies that sorcery implies. While Genaro is, according to himself, not a person who initiates witchcraft, he certainly punishes people through witchcraft on behest of his clients. In his testimony, he not only outlined some of the calendric practices in witchcraft, but also made me visualize how he attacks wrongdoers and initiators of witchcraft. Thus, witchcraft is defined, on his side of things, through a series of practices: the identification of a victim, the performing of ceremonies at appropriate calendrical occasions in the graveyard, the invocation of the wild *win* spirits or the souls of the dead to summon, the finding of an enemy through the 'blood-speak' and sending the spirit against him. He has a patron, like other sorcerers do. A cosmology sustains his practice, quite similar to that presented in Alberto's testimony. However, besides these details, the interesting thing is the emphasis on the encounter of the gaze of his enemy with his servant spirits: 'I go to the cemetery to call the dead before my enemy, so he sees the dead in the night, so he is frightened. The snake will appear to him in the street. I summon the snake, a spiritual one, and it goes through the air.' This encounter also implies some sort of temporal dimension to it, for the illness and death only comes after it: 'Let's say he is lying on his bed. He sees the snake in the night. He is frightened: the illness comes … He steps on his grave.' Yes, the man sees the snake spirit or the dead somewhere, but then he starts seeing them at other places, getting sick, etc. Thus, the initial moment, the gaze on the unnerving, the dazzling, is needed, as well as a development to it, a temporal dimension when the sorcery as a series of 'attacks' starts to develop.

Furthermore, the interruption of the call from the man at the beginning of my interview can give us an almost synchronic image of the extended causality of sorcery, that which Evans-Pritchard evokes in his account on the emergence of *why* questions in witchcraft. Basically, the victim commits an offence or is simply hated by someone, and then the sorcerer makes his working, perhaps on behest of a client, or even of himself, as Genaro explained. Then, from the sorcerer's point of view, he sends a spirit to attack his victim. This spirit is then 'seen' by the victim, perhaps in a direct way or in an indirect fashion, for they only see its effect: an accident, bad luck or social failure. What I consider crucial is to consider how the intentions of the sorcerer and the perception of the victim regarding causality somehow 'collide', perhaps a way akin to the idea of 'collision of paths' worked out by Holbraad in regards to Afro-Cuban divination (2012: 224–33).

Discussion: Witchcraft as a Usurpation of Causality

The testimonies in this chapter seem somewhat disjoined; after all, I could not produce out of my field material a consistent ethnography of witchcraft, as the ones that characterize the works mentioned in my introduction (Favret-Saada 1980; Kapferer 2002; West 2005). However, I believe that, despite this, a certain intervention in the discussion of this topic can be made by comparing all the testimonies presented in this chapter as describing different facets of the same phenomenon from the point of view of a ritual specialist (Alberto), a practitioner of counter-sorcery (Genaro) who, nonetheless, can damage people with his magic power, and an indirect victim (the man on the phone).

The testimony of Alberto talks about an asymmetry of power, the contrast between the assertion 'I can harm you' versus the other assertion of 'you can do nothing to me'. Thus, it implies a logic of attack, but also a hierarchy of causation: the sorcerer can influence the victim, but not the opposite, at least at first. The testimony of Genaro describes *how* this attack occurs: the sorcerer goes, does his ritual work, which is abetted by powerful cosmological patrons and other spirits, and then a person 'finds' the spirit in the form of any accident or happening. Now, the impromptu call of the man in the testimony of Genaro is of great importance, because he exemplifies how victims perceive sorcery, or, more precisely, how the idea of sorcery arises in the eyes of the victim. The man had an accident, and suddenly he thought to himself that the accident was not naturally caused, but caused by sorcery. Now, I want to delve into this point a bit more in a theoretical fashion.

Of course, the episode of the man having an accident and immediately thinking about witchcraft as the real cause behind it is rather similar to the famous account of Zande sorcery by Evans-Pritchard (1976: 22), alluded to in the introduction to this chapter. As I mentioned, recently Martin Holbraad suggested a way to analyse Pritchard's insight in terms of 'how' versus 'why' questions or logics (cfr. 2020). In the case of the Zande man, the question was not 'how' the roof of the beer hut caught fire, but 'why'. Thus, the 'how' is the causal chain of events that happened, by the 'why' is something different. Holbraad characterizes this as a contrast between causal and non-causal interactions, and relates this to his proposal regarding colliding trajectories of motions in Afro-Cuban divination:

> The chains of causation that how-questions rely on start with discrete events (this happened, that happened, and so on) and then set them into motion by connecting them sequentially, like a cascade of dominoes. Inversely, the non-causal intersections that why-questions involve begin with vectors of motion (namely, the unfolding of two separate causal chains) from which a discrete event is derived, by way of collision (namely, the moment of "coincidence" Evans-Pritchard invokes). In short, the "how" starts with discrete events and ends up with moving trajectories, while the "why" starts with trajectories of motion and ends up with discrete events. (Holbraad 2020: 507)

And to this, the sorcerer will say something akin to this: 'The *why* is nothing else but my actions. I did it. The accident or misfortune that befell you was my doing. I sent [for example] a snake flying through the air, and it met my victim, and it harmed it. It is my power.' Of course, this does not 'make sense' to us; how is this caused by you? But in a way, this assertion is similar to many other assertions throughout this book: it is not only people who 'do things', but gods or cosmological entities.

What I can propose now is this: sorcery is very similar to other cases of cosmological causation presented in this book; that is, causation as perceived as flowing from the 'above' of cosmology, rather than from the normal sequence of cause and events that characterize our world. However, it is also similar in the dilemma that, as we have seen, many times cosmological events are actually caused by people and their ritual actions, rather than by gods. What we see is that, in witchcraft, the sorcerer is like a usurper of causality. People experience misfortunes, many times in an actual sequence, and instead of attributing them to the 'how' of normal causation, they attribute them to the 'why' of sorcery. My argument, however, is that this is still a form of causation, but it is a sort of 'hijacked' causation that flows from the sorcerer to the victim, rather than the normal sort of cosmological causation that characterizes daily life. This is why the idea of asymmetry in the first testimony is so important. The relationship between the sentences 'I can harm you' and 'you can do nothing to me' is the essence of sorcery, at least in its initial encounters. It is also important to see the contrast between the 'observed' and the 'observer' in it. Just as Alberto reversed his condition of being a mere 'object' of the ethnographic gaze and became the subject of another more powerful counter-gaze, in sorcery the normal flow of things is reversed by a more powerful level of causality that assumes direct responsibility over some of the happenings of my life. Thus, what I argue is that these happenings are not 'non-causal' in nature, but are retransformed into being caused in terms of a superior instance that from now on redirects them or constrains them according to its needs.

Returning to the collection of testimonies, of course witchcraft does not end here. Indeed, the sorcerer can cause me harm in a way that I cannot 'act' back in a normal way, as is possible in any other social interaction. What I need to do is what the people in Momostenango usually do in these cases: perform counter-sorcery.[8] This is a restoration of 'causal' powers to the victim, a reinvestment with agency. Again, what I consider is that sorcery can be conceived as a hierarchical layering of causality, rather than a negation; that is, the emergence of a malicious causal agent who is in a relationship of hierarchical superiority to that of the victim, usurping the usual flow of cosmologic causation. But with the fighting back of the victim, things get even again, and the circuit of sorcery can be eventually closed (fully or only partially), as Favret-Saada famously established (1980: 194–222).

Notes

1. Cleverly, West notices that all anthropological theorizations can indeed be seen as ethnographic sorceries themselves (2007: 83).
2. 'It is worth remarking, however, that special knowledge that inheres, say, in theological or scientific expertise, has never held quite the place in anthropological accounts as materials that appear esoteric *because* they require revealing (beg immediate interpretation). An initial surprise becomes a suspension, a dazzle, and some kinds of "special knowledge" are more likely to dazzle than others. One is held, as it were, on the threshold of understanding. I referred to having been mesmerised' (Strathern 1999: 10–11).

3. The *win* is a spirit, either that of an animal or a dead person, which is used by K'iche' sorcerers in order to harm their victims. In some versions, as the testimony of the sorcerer in this chapter, this spirit is merely summoned or asked from a *dueño* or patron. In other versions, the *win* is a transformation of the sorcerer's soul itself, which 'flies out' of the sorcerer's mouth after performing some ceremonies and doing nine backward somersaults. In those versions, the sorcerer himself is disguised as a *win*; thus, the word *win* denotes both the sorcerer and the spirit (cfr. Zamora Corona 2019: 124–9).
4. The seventeenth day of the K'iche' calendar. The word in K'iche' means 'thought', but the meaning in both the Aztec and the ancient Maya versions were related to the earth: the Aztec equivalent was *olin*, 'earthquake', and the Maya *kaban*, *k'aab'* meaning 'earth'.
5. Similar legends exist among the Tzeltal Maya of Chiapas, where certain mountains and places of sorcery exist, where people can sell souls to the lords of the mountains or *dueños* so they become indentured servants of the *dueños*, looking for their animals (Pitarch 2010: 35).
6. I have seen such altars, although not 'in the flesh', but in the photographs of a colleague.
7. This testimony is interesting, for it somewhat corresponds to Alberto's information that the deer of the underworld are the souls of dead sorcerers; Genaro only would insist that it is the soul of a dead man.
8. See Zamora Corona 2021: 195–201 for more testimonies of sorcery in Momostenango, in which counter-sorcery plays an important part, and that adjust more to the model of Favret-Saada; that is, counter-sorcery as a mere redirection of the evil power of the sorcerer back to himself. In the case of Genaro in particular, he does not simply redirect the evil intentions of the original sorcerer to his but, rather, he uses *his* own sorcerous power to punish or even destroy him.

5

The Struggle for the Earth

The Cosmology of Activism and Resistance

> They do not consult any medical man; here on the mountains, where people still hold by ancient ways and customs, they go to a wizard. Each tribe, distinguished from the rest by speech and dress, has at least four of these aged wiseacres. By means of charm and spell they drive out all manner of evil spirits, which are supposed to be the cause of all complaints … These wizards also exercise some sort of authority, acting as headmen of their villages … They keep a careful watch over the mineral treasures of the mountains. It is known, for instance, that there are rich quicksilver deposits, but if any white man attempts to start prospecting, the wizard sends for his satellites, and the expedition never returns. Many things may happen among these mountain ravines which are never brought to light.
>
> —Prince William of Sweden, *Between Two Continents, Notes from a Journey in Central America*

The quote that figures at the head of this chapter (Wilhelm Ludwig 1922: 215–16) was written by Prince Wilhelm, Duke of Södermanland, alluding to the trip he made to Guatemala and other countries of Central America during the final years of the dictatorship of Manuel Estrada Cabrera, who ruled the country from 1898 to 1920. The anecdote itself does not refer to Momostenango, but to the nearby K'iche' town of Almolonga, near Quetzaltenango; however, it shows that the current-day struggle of the Maya peoples of Guatemala against modern Western extractivism has a rather long history.[1] The 'wizards' mentioned by Prince Wilhelm are nothing but *ajq'iij* or K'iche' calendrical specialists who, just as today, held important appointments in the Indigenous government of their towns. However, this is not the only continuity that this quote implies. In fact, just as in the past, the struggle against mining companies is not only grounded in local politics but it is also grounded in a close relationship between cosmology and power. The passing and rather exoticizing observations of Prince Wilhelm do not explain why the K'iche' people of his day kept watch over the mineral treasures of the mountains, making it sound as if it was, perhaps, only related to the assertion of Indigenous control over their own territory. The truth, of course, is more complex, and it has certainly to do with the cosmological conceptions described in this work.

This chapter will present the story of one of the most politically engaged and ritually accomplished among my informants: Miguel Ángel Pelicó, a K'iche' activist widely regarded as a strong *ajq'iij* in Momostenango. Miguel Ángel is a member of the Autoridades Ancestrales movement, a movement of revitalization of Indigenous political and social practices and resistance against extractivism and other threats to Indigenous territories. As I hope to show in this chapter, this kind of activism is not merely nativist or 'neo-Maya' in its scope or ideas, as it has been characterized in former works (cfr. Ponciano 2007). Instead, I will show that what this movement has done is to follow the political structures and the cosmological logic extant at the communal or 'traditional' level, taking them to a new level where national and global issues can be addressed.

Regarding its theoretical outlook, this chapter will engage with a subject that is currently in vogue in anthropology, that of what has been called 'cosmopolitics', in the sense of Isabelle Stengers (2005, 2010), retaken by Marisol de la Cadena in her work about Runakuna (Quechua) shamans and political activists, *Earth Beings* (2015). The topic, of course, is not new, and it could be said to stem from what has been called 'politics of nature' (Latour 2004), and is certainly adjacent to the work of anthropological and political theorists like Elizabeth Povinelli (2016, 2021). Indigenous struggles against Western extractivism have been considered as a strong starting point for a wider debate on ontology, politics and ecology (cfr. Questa 2023; Rivera Andía and Vindal Ødegaard 2019: 6–13); however, there is no lack of controversy regarding how the complexity favoured by the ontological school sometimes seems to overlook the everyday life groundings of social and ecological practices (Killick 2015: 15). While ontological approaches have a potential to create a different kind of politics (Hage 2012), more ethnographic studies on Indigenous resistances to extractivism can cast more light on this and other debates (Rivera Andía and Vindal Ødegaard 2019: 10).

Indeed, another kind of criticism can be levelled against some recent examples of ontological approaches, which does not completely detract from the general aims of this 'school'. In general, recent ethnographic accounts of Indigenous resistance against extractivism emphasize a conflict of ontological differences, presenting the 'cosmopolitical' conflict of Western extractivism versus the Indigenous 'politics of nature'; usually, the outlook is, justifiably, politically pessimistic. While cosmopolitics has always emphasized the place of Indigenous agents of resistance, as De la Cadena's account has, something that is missing from some cosmopolitical ethnographies is precisely how Indigenous practices articulate resistance, not only in terms of being ontologically opposite to Western extractivist agendas, but in how the ontology that emerges from these practices *empowers* Indigenous agents in order to resist these agendas and become wielders of cosmological and political power themselves, both inside and outside their communities. Thus, the aim of this chapter is to present a side often overlooked in this story, that of the Indigenous practice of power, and, furthermore, to present a story of successful Indigenous resistance to extractivism.

Ancestral Authorities: Indigenous Activism and New Political Structures

The political organization of Momostenango has naturally changed since Robert Carmack described it on his classical monograph on the topic (cfr. 1995: 310), which reflects the conditions prevalent prior to the Peace Accords of 1996, a watershed moment in Guatemala´s history. In Carmack's account, the authority structure of the town could be divided in two: a Ladino block and an Indigenous block (Carmack 1995: 309–28). However, the authority of the Ladinos has greatly declined since then. A Ladino lady in town told me succinctly: 'We Ladinos are finished' (*Nosotros los ladinos estamos acabados*). Nowadays, according to my informants, political power can be divided into three kinds, all Indigenous, the first two being the communal[2] and municipal[3] authorities, which could be compared to the 'traditional authorities' and the 'local officials' described by Carmack (1995: 310). However, there is a third kind of authority, of relatively recent creation: the so-called 'ancestral authorities' (*autoridades ancestrales*), an emerging movement that has taken root among all Indigenous groups in Guatemala, not only Maya,[4] and that can be described as being an activist movement in its scope, understanding the latter as movements that 'challenge some existing element of the social or political system' (Saunders 2022: 1).

The self-described mission of ancestral authorities is the resolution of social and political conflicts according to Indigenous tradition (Arifín-Cabo 2011: 2). Some informants in Momostenango described them to me as 'cultural', a correct assessment regarding the origin of their conception of power, for the basis of its existence is the notion that there is a difference between the fact that a political authority is Indigenous and the fact that this authority is actually considered 'ancestral'; that is, 'truly motivated' by Indigenous tradition (Arifín-Cabo 2011: 7–8). This assertion seems contradictory when we ponder on the fact that communal authorities in Momostenango are indeed 'traditional'; however, ancestral authorities tend to reject the system of *alcaldes* and the like as creations of the colonial period (given, among other things, their usage of Spanish colonial terminology) and instead appeal to a certain notion of Indigenous authenticity; thus, the notion of ancestral authority is founded on a conception of power born from 'nativist' activism and, specifically, from the so-called 'Maya spirituality' movement.[5]

Ancestral authorities throughout Guatemala organize themselves in different councils or *consejos* integrated by respected members of Indigenous communities that engage in traditional practices and are regarded as 'Maya authorities', such as calendric priests (*ajq'iijab'*), Indigenous magistrates (*q'atal tzij*) and midwives (*ajkun*), just to quote a few examples of likely candidates (Arifín-Cabo 2011: 8; SERJUS 2014). In practice, they function exactly like the *pasados* system of the communal authorities, which means that one can only become a member by being appointed by former members (*pasados*) of these councils. In the case of Momostenango, the relevant organization is the Council of Ancestral Authorities of the 31 Communities of Momostenango (*Consejo de Autoridades Ancestrales de las 31 Comunidades de Momostenango*), 160 traditional authorities who gather two or three times a year. Miguel Ángel offered me its basic outline, which is very close to the scheme of communal authorities. The Council is integrated by nine members who rotate every two years.

A few words on the conception of power that is at the basis of the movement of ancestral authorities is needed here. An interesting summary of the critiques that this kind of organization levels against other Indigenous forms of authority can be found in a document prepared for the German Corporation for International Cooperation (*Deutsche Gesellschaft für Internationale Zusammenarbeit*) by Pressia Arifin-Cabo, Deputy Representative for UNICEF:

> Current day Indigenous mayors (*alcaldes*) can barely be considered traditional authorities because the appointment of an *alcalde* doesn't occur through traditional procedures. The position of *alcalde* and Indigenous *alcalde* were

> created and imposed by the Spaniards. Indigenous authorities can be elected by popular vote or by official appointment, while ancestral authorities are elected on the basis of their *nawal* or energy. The conception of authority is concentrated on a person that serves the community and works for it, not on a person that holds power over the community. (Arifín-Cabo 2011: 7)

Thus, ancestral authorities criticize regular Indigenous authorities not only on the basis of their colonial historical roots, but also on the basis of an idea of power that is defined as a service to the community, an idea that closely resembles the slogan of an organization like the Mexican Zapatistas, '*mandar obedeciendo*' ('ruling by abiding', cfr. EZLN 1994). Finally, we need to stress that the political mission of these groups goes beyond the solution of conflicts on the basis of an authority that responds to Indigenous interests and emanates from Indigenous conceptions. At least four specific causes, all of which overlap with the agenda of the Ancestral Authority council in Momostenango, are stated as crucial for them (cfr. Serjus 2012):

1) The defence of Indigenous territories (particularly against private extractivist projects that are closely aided by the neoliberal Guatemala state).
2) The recognition of Indigenous law and jurisprudence.
3) Demanding more accountability from local authorities.
4) Proposing alternative, sustainable plans for local development.

As we will see, the first and the third cause are the most immediately relevant for our chapter, especially in the life of Miguel Ángel Pelicó and other leaders; however, Miguel Ángel and other members of the *ancestrales* movement have regularly pushed the rest of the points of the agenda, signing declarations of law reforms in Totonicapán with the aim of reappraising native law systems in Guatemala (cfr. Consejo Departamental de Autoridades Ancestrales y Comunitarias 2016).

Miguel Ángel Pelicó: Local Authority and Activist

At first, during my fieldwork, I wanted to concentrate on a *costumbrista* family such as the Ramos family, but I was also curious about Indigenous politics of resistance against multinational mining projects, which, as I have mentioned, are a strong subject of conflict in present-day Indigenous Guatemala. Before I arrived in the town, I already knew that, in 2013, there was a referendum organized in town rejecting mining companies, a huge victory for Indigenous territorial rights in Momostenango (Domínguez 2013). However, while also sympathetic to the cause, Virgilio and his family were not involved that much in its organization. Instead, this specific movement of resistance was organized by the Council of Ancestral Authorities, with the help of other Indigenous organizations. Thus, I always kept an eye open for the chance to meet someone involved in these activities.

The opportunity came a few months into my fieldwork, when looking out for calendrical specialists to interview, one of which was Miguel Ángel Pelicó, who recently had consecrated a health clinic built in town by a foreign non-governmental organization.[6] I was invited to his home to interview him, and this meeting would become the origin of a very fructiferous field relationship. The day of our meeting, after arriving at his home, his wife greeted me and conducted me to the family's dining room, where Miguel Ángel appeared. A serious-looking man in his mid-forties, Miguel Ángel has a certain aura of authority and command (Figure 5.1.). He then told me that he missed the first appointment because he is an electrician and he had to check electricity meters all around the town, and he had a particularly bad day at work. In fact, Miguel Ángel owns a motorbike, which he skilfully rides through the town, whose mountain roads are particularly steep and treacherous due to lack of maintenance.

Miguel Ángel then explained that, on that very day, he also had to consecrate a *meb'il* stone (we met on a 10 Tz'ikin day, a very propitious day to carry out such a ceremony), but he had time to reply to a few questions. He reintroduced himself, stating his calendric *nawal*, the day he was born: 'I am 13 Kiej. People obey me, for that is my *nawal*. When I speak, everyone obeys. This doesn't happen because of me, nor because of what I am. This is what God wants.' Indeed, in K'iche' calendrics, Kiej is the day related to political power, being one of the days that can be *alcaldes* of the Mundo,[7] and thirteen is the highest coefficient in the divinatory calendar, usually implying accumulated force, so it is a very impressive *nawal* to have. Miguel Ángel attributes his domineering nature in his daily life to his *nawal*: 'If I tell my son to come home when he is late, he obeys. My *nawal* is commanding, it is strong.' Then, Miguel Ángel proceeded to explain a bit about how he became a calendric priest and a traditional authority.

It all began at the end of the nineties. At the time, he was not a traditional lineage head, nor a calendric priest. He did not inherit any *awas* shrines. However, he had a 'shamanic calling' of sorts. He had bad luck

Figure 5.1. Miguel Ángel Pelicó burning offerings during the calendrical ceremony of Wajxaqib' B'atz'. © Alonso Zamora Corona

and illness, and he went to a divining priest, who told him that his soul was a prisoner in the underworld. His soul was then freed by him, and Miguel Ángel was given the opportunity to become a calendric priest himself; according to the calendric priest, his illness was precisely the result of him not attending to his call. Therefore, Miguel Ángel sought initiation as a calendric priest, and this opened to him access to the communal authorities of his ward. Thus, he became the traditional *alcalde* of Barrio Santa Ana, one of the main wards of Momostenango. He also became involved in the ancestral authorities movement and was appointed as the president of a side organization that helps the movement, the Coordinadora de Organizaciones Indígenas de Momostenango (COIMO). According to Miguel Ángel, it was all because of the force of his *nawal*; he said the *pasados* liked him quite a lot, because of his strong personality. While asked about the relationship between the native calendar, organized on political lines, and himself, Miguel said, 'I am an authority: people listen to me, hear me, obey me, for I am a spiritual authority too: I can rescue souls from the jail of the earth gods. I rescue people, for I am a *licenciado* before the Santo Mundo'. Thus, he conceives his position as a calendric priest as a sort of political or bureaucratical dealing with the underworld, which, at the same time, strengthens his position in actual politics.

Miguel Ángel then expounded on his political vision, which is the one of the ancestral authorities movement. His political cause is 'the defence of our territory'. Furthermore, he is engaged in a more nativist view of *costumbrismo*, rejecting some of its Catholic elements. He explained that the memory of the conflict between Acción Católica and *costumbre* is still alive, and he recalled some episodes that happened in the pre-civil war period, where members of Acción Católica would destroy ancestral altars and ritual paraphernalia.[8] He compared these acts with the persecution of native religion during colonial times: 'As for ourselves, they burned our legacy, our knowledge. What was the consequence? Many were blinded, became deaf, lost their hands, fell on rivers, under mudslides, the lightning strikes, disgrace came, darkness, our ancestors suffered it, we were enslaved and persecuted.' This rejection even stretches to the language he uses in his prayers, a subject on which we will come back: 'I only use pure K'iche'. Not a single word of Spanish in it.' He certainly offers cult status to the saints, uses crosses and the Bible, and prays at the main Catholic church in Momostenango, but he is careful to emphasize the K'iche' elements of the ritual over the Catholic ones. However, as I will explain later, this purism, which is not of the extreme variety, is not at all rejected or perceived as artificial by *costumbristas*.

Miguel Ángel also relies on the divining power of the 'blood-speak' mentioned in Chapter 1 to make personal assessments. At one point during our first meeting, he even started to divine about my visit (I did not ask him to), counting days with his hand and stopping whenever 'his blood' told him to. 'Aq'ab'al, a new beginning, the dawn. Tz'ikin, a fortunate day. Kawok, you will march slowly. Tz'ikin, good fortune prevails.' Thus, he concluded that our encounter would somehow 'bring a new beginning', and that I would be lucky in my endeavours. While this would seem like a gratuitous detail, I need to stress that many of my interactions with my *ajq'iij* informants were dependant on how they perceived my intentions based on what their blood told them about me. Sometimes I was lucky, and I was trusted; in fact, I made a lasting friendship with Miguel Ángel, who always granted my requests. But on some occasions, I often wondered whether a rejection without apparent cause had to do with a negative divination verdict.

At the end of the interview, Miguel Ángel opened a huge curtain at the back of his dining room. The curtain separates Miguel Ángel's altar from the room. It is not a stone altar like those of Virgilio and Rudy, but a table with ceremonial implements. Decorations on Virgilio's table altar were sparse, while Miguel Ángel's was more spectacular, with a huge Bible, a huge icon of the Mexican Virgen de Guadalupe, an image of Christ, of Saint Michael, as well as K'iche' clay figurines, representing the days of the calendar personified. I will not deal here in detail with Miguel Ángel's ceremonial activities, which I frequently attended. However, I must stress that his prayer style was well liked, for it was not really a neo-Maya style of prayer (which incorporates many New Age fabrications), but more of an old-school style of prayer, mainly based on old expressions in K'iche' and excluding Spanish terms. In fact, many of the ritual addressing forms that I mention throughout this book come from his prayers, but their legitimacy is confirmed by the fact that they have been previously recorded in the old ethnographies of Schultze-Jena (1933). When working on their translation, my Protestant friend and K'iche' teacher Abraham praised him vividly: 'This is pure K'iche" (*esto es K'iche' puro*), he said, and then he clarified that his usage was really elegant. Even as a Protestant, Abraham praised the usage of authentic expressions of addressing the ancestors, and clarified that K'iche' language is considered to have a particular quality to it, *q'aq'al* or 'heat', not unlike the notion of *nawal* explained in previous chapters, a sort of description of its capacity of rhetoric persuasion; in contrast, Spanish is considered to be 'cold', to have a weak *nawal* and thus it is not as useful as sacred speech, even to the point that Protestants favour the recitation of K'iche' versions of the Bible, and even 'feel' the *nawal* of K'iche' language raising as 'heat' during prayers. Thus, he recognized that Miguel Ángel's style and language was peculiarly moving, while conveying an idea of command and authority, not unlike the idea of *nawal* as the heat-force of someone's co-essence. He literally said to me about this particular language: *Are la' usumil ranima* ('This is the deepest recess of our souls').

Miguel Ángel's Career as an Activist: COIMO and the Ancestral Authority Council

Nowadays, Miguel Ángel has mostly ceased being an activist, concerning himself with his private life. This is not out of a loss of conviction in the movement or connection with it, since he regularly attends meetings, but because, by the principle of authority rotation, new people have come to the forefront and also, after his personal victory over the municipal *alcalde*'s tax plans and the collective victory over mining concessions, Miguel Ángel's two main political objectives were (at least temporarily) achieved. Nowadays, the Council of Ancestral Authorities in Momostenango concerns itself with pushing other agendas, like the recognition of Indigenous jurisdiction and its sustainable development agenda. Thus, he would usually characterize the situation by saying: 'We are sleeping now, but we are always vigilant.'

Miguel Ángel and I would usually go over his activist career during interviews at his home. His extensive political career began after he became a calendric divining priest. As I have mentioned, at some point he entered into the Communal Authority system, becoming the *alcalde* of the Santa Isabel ward, one of the four main wards of the Momostecan town centre. He was introduced to the ancestral authorities movement through similarly minded prominent Momosteco activists, like the ex-guerilla fighter and activist Abraham Vicente, the former leader of the Council of Ancestral Authorities or *nab'e kamalb'e* ('first guide') at the time of the antimining referendum; and Gumersindo Kalel, and his friend, the late David Coxaj, who was *nab'e kamalb'e* at a later point and passed away in 2018. Eventually, he was appointed as the president of COIMO in 2011,[9] and served as either first or second president (*kamal b'e*) of the Ancestral Authorities council over four years (2012–2015). As leader of COIMO and prominent member of the Ancestral Authorities council, Miguel Ángel pursued a general agenda of demanding accountability from the *alcalde* at the time, Santiago Federico Pelicó Rojas, from the UNE party, and his municipal corporation, who would usually seek to implement decisions without taking into account public opinion, while using political connections to dominate and coerce the communitarian or traditional authorities.

Miguel Ángel's relationship with the town *alcalde* was rocky from the start. As *pasado* of the Santa Isabel ward, president of COIMO and member of the Ancestral Authorities council, Miguel Ángel had great influence in the communal resistance movement, and his pursuit of municipal accountability saw him at odds with the municipal *alcalde* on many occasions. On 4 November 2011, Miguel Ángel was violently attacked on the road to his home near Pak'lom Hill, by seven hooded men who were riding motorcycles, while he was strolling around with a companion. It was an act of intimidation, for he was not robbed, and his companion was left untouched. He was beaten up and many death threats were made against him. Fortunately, his companion came back, feigning that he had gathered some help, and the attackers fled. The news report on the incident at the time stressed that this was due to Miguel Ángel's political activity (CERIGUA 2011).

Miguel Ángel himself attributes the attack to the municipal government under the *alcalde*. Politically, but ritually too, these events confirmed for him the necessity of having a new kind of Indigenous authority oriented to the defence of the territorial claims of the community, beyond the municipal/communitarian government dichotomy. For Miguel Ángel, his career as an activist and politician was a time of intense ritual activity, not only because he had to carry out rituals to aid the success of his political dealings, but also because he carried out rituals to protect himself from political sorcery. Indeed, the *alcalde* had his own personal calendric priests, who, in Miguel Ángel's view, carried out witchcraft against him. He says that there were other attempts on his life that never made it into the press. Ultimately, he attributed his repeated survival to the strength of his *nawal*. Discussing these events, he said to me: 'Nobody can kill me, even if they beat me up: my *nawal* stops their bullets. They fear me, because they cannot kill me.'

And indeed, despite these physical attacks, Miguel Ángel did not back down. A major, final victory for his administration at COIMO was achieved in his successful organizing of the Indigenous resistance against the real estate tax known as IUSU (Impuesto Único Sobre Inmuebles; see SAT 2012). The implementation of this tax, of general application in Guatemala, was attempted in 1997, just after the civil war (1982–1996), but local resistance achieved the derogation of it in Totonicapán, the Guatemalan department to which Momostenango belongs administratively. In 2015, the mayor of the town, Santiago Pelicó, wanted to reimplement the IUSU, which prompted widespread resistance in the town.

On 26 June 2016, COIMO, with Miguel Ángel at its head, convoked a public demonstration before the municipal building in the town centre, demanding the mayor's resignation. Four thousand people participated, and this manifestation was supported by the Ancestral Authorities council (Domínguez 2015a).

In retaliation, the mayor sued Miguel Ángel and all the members of the Ancestral Authorities council for 'coercion' (Batz 2015), calling them before the judiciary authorities of Totonicapán. However, shortly after, the authorities dismissed the plaint and the IUSU plans were scrapped, which prompted the reaching of an agreement between the municipal *alcalde* and the *ancestrales* on 1 July (Domínguez 2015b). This was celebrated as a big triumph for COIMO and the Council of Ancestral Authorities. Furthermore, they even countersued the *alcalde* in August 2016; the plaint was dismissed too, but the outcome was considered as a clear proof of the power of the organization (Domínguez 2016).

Regarding his usage of the calendar in his political dealings, Miguel Ángel told me that there are days for protection, and days useful for political dealings. 'Imox makes you invisible, so you won't be found, but also to reveal an enemy. This is what the Momosteco soldiers used in the war against El Salvador … Tijax ['flint knife'] cuts the slanders of your enemy. It is used to cut evil, evil intentions and evil acts.' Furthermore, he equated his authority with this calendric *nawal*, for the word for an authority in K'iche' is *q'atb'al tzij* 'the cutting word': 'My word has authority, because it cuts, with Tijax I cut the words of others and impose my opinion.' Other days were used to reach agreements, like Kiej ('deer'), which, as we have seen in previous chapters, is used by the communitarian authorities but also by the Ancestral Authorities council for their meetings. B'atz ('thread') is used to wrap up things, like an agreement: he performed such a ceremony after the referendum. As we have seen, though, it is not really a conflict of *costumbristas* versus Ladinos anymore, but one of K'iche' vs K'iche', and thus both sides resort to ritual practices to achieve their aims. Miguel Ángel says about this: 'My enemies do burn offerings too, but they are sorcerers.' Thus, Miguel Ángel's political activity had in his eyes two facets: one as a communitarian leader, the other as a cosmological *licenciado*, propitiating the earth and calendric gods in order to achieve his political objectives; in fact, the word he used to characterize his rituals was *despacho*.[10] The word means 'remittance' of an offering, but in Momostenango it also alludes to an actual audience in the 'office' (*despacho*) of a calendrical/earth divinity, which are seen as bureaucrats of the Santo Mundo or 'Holy Earth'. In fact, Miguel Ángel's ritual language was always evocative of such a gathering, describing how the *dioses mundo* entered his altar-room to hear his petitions (Table 5.1).

Table 5.1. Opening of a ritual prayer in K'iche' Maya, and its English translation

Rí ajaw, e remelik, e tikilik e chemelik.	The lords are sitting, gathered and present
Wane na'bi ch'ob'olik	Maybe there is no refinement
Ru le mik ru nutik chwachalaq	In my words before you,[11]
We ne kinb'e na q'iij,	I may take too much of your time,
Wetam taj.	I don't know.
Kin jach chiwach alaq	I bequeath in your presence
K'olik ri nu kotz'i'j,	These flowers of mine that are here,
Ri nu chaj.	These candles.
Kin b'ik che alaq:	I tell you:
K'ama alaq, taka' alaq,	Receive them, hear them
K'ama alaq, taka' alaq,	Receive them, hear them
Ri tzij, ri k'olik …	The words that I say before you …

Usually, after a similar introduction, he would expose his 'case' before the earth divinities. Sadly, I could not record a prayer with a political content: most of them were concerned more with the worries of his current clients on divination. However, Miguel told me that vying for the soul of someone in the underworld consisted of an exercise of rhetorical persuasion with these entities to side with him, trying to win them over, as in a legal battle. Thus, it is not hard to imagine a similar cosmological 'legal battle' between him and his enemies taking place during his political dealings, each one of them burning offerings to further their cause. When I asked him about the parallels between the calendrical order, which has a hierarchy, the so-called 'government of the earth' (*gobierno mundo*) and the town's political order, he told me: 'This is the order of things: there is an order in town, and there is an order in the *Mundo* too.' What is interesting is how the 'doors' to access both powers are somehow interconnected, but I will return to this in my conclusion. For now, I want to finish my account on Miguel Ángel's activities, in particular regarding his participation in the resistance against mining companies.

The Struggle Against Mining Companies

Approaching the end of my fieldwork, Miguel Ángel invited me to the celebration of the anniversary of the referendum against mining companies, a huge victory for his organization. The referendum took place on 1 September 2013, and thus this was its fifth anniversary in 2018, when I was in town. However, before describing the commemoration, some context is needed.

It is a badly kept secret that the mountains of rural Momostenango probably hide veins of gold and silver, as well as other ores. According to my informants, this has been known for many years, even before the beginning of the twentieth century.[12] While not being formerly recognized as a region particularly endowed with mineral resources (Carmack 1995: 13; McBryde 1947), similar information has encouraged mining companies to attempt exploration and prospecting activities. In 2001, the government of Guatemala granted permission to a mining company, Montana Exploradora de Guatemala, part of the Canadian mining conglomerate GoldCorp, to carry out prospecting activities in Momostenango, without the permission of the inhabitants of the town, although with the probable approval of the town's mayor. It also granted numerous mining concessions to the company. The main project was called Centauro II, and it contemplated the mining of gold, nickel, silver, cobalt, chrome, antimony, zinc and rare earth materials. (Observatorio de Conflictos Mineros de América Latina 2024).[13]

While not being formally started, the prospecting authorization was renewed in 2007 and again in 2012, something that raised great indignation in the community. However, the mayor of Momostenango at the time, Santiago Federico Pelicó Rojas (the same *alcalde* with which Miguel Ángel had the tax conflict), was secretly partial to mining and, while saying in public that he was opposed to mining projects, he himself arranged the renewal and granting of permits at Totonicapán and Guatemala City, which prompted the beginning of the movement that was organized both by disaffected communitarian authorities and ancestral authorities. The three figureheads of the movement were Renato Vicente, Gumersindo Kalel and Abraham Vicente, a veteran guerrilla fighter during the civil war. It began with a legal protection appeal (*amparo*) presented by the communitarian and ancestral communities before the Constitutional Court of Guatemala. The reason of this appeal was the absence of a legal local referendum regarding mining, which was a violation of the ancestral territorial rights of the Momosteco K'iche' and of the Indigenous and Tribal Peoples Convention of 1989 (General Conference of the International Labour Organisation 1989), which the Guatemalan government ratified. The court decision was favourable to the communities, but the Ministry of Energy and Mining appealed against this decision, which prompted further litigation. In the context of this litigation, the communal referendum was organized and conducted on 1 September 2013 (Domínguez 2013).

The content of this complaint was not only regarding territorial rights conceived as property, but it had a dimension related to K'iche' religious life, explicitly recognized in the communities' plaint: 'The communities were concerned about the impact that mining activity will have on K'iche' spirituality because the mountains that are projected to be mined have a fundamental function in the town, where sacred places exist' (Observatorio de Conflictos Mineros de América Latin 2024). Similarly, a crucial aspect was the environmental impact of mining activity, which could lead to extensive contamination and damage to the community. Finally, a municipio-wide referendum was organized by both the Council of Ancestral Authorities and concerned communitarian authorities. More than 50,000 people voted overwhelmingly against mining. The referendum gave more strength to the ancestral authorities' argument before the constitutional court, and thus, in 2015 the court decided against the Mining Ministry again.

As president of COIMO, Miguel Ángel was close to the movement, and on one occasion he showed me some of the locally produced DVDs with their reunions, as well as the referendum itself: the discussions and speeches were carried out mostly in K'iche'. The video of the election of the ancestral authorities belonged to the post-referendum period, being filmed on 19 January 2014, but it was important as a depiction of the strategies and aims of the organization after their victory (Cámara Nueva Visión 2014). The Council of Ancestral Authorities always timed its meetings, like their elections of new authorities, to be celebrated on days associated with political power, like Kiej: this fact was mentioned during the meeting. The referendum itself was conducted on an 11 Ix day, associated with the earth, and given its high numeric coefficient, particularly strong. During the election of 2014, Kalel made traditional appeals to the ancestors, and the council had a K'iche' calendric priest who prayed before them, appealing to the dead ancestors in the same ritual language used by Rudy and Miguel Ángel: '*qanan qatat, sak ki wi, sak ki jolom*' (mothers, fathers, white are your heads and hairs …). Thus, ritually and politically, their meetings were somewhat reflections of the ones made at the communal level. A gathering of the members of the council deliberates and nominates the new members of the council, who receive metal-tipped colonial *varas* just as the communal *alcaldes* do. During the meeting, long speeches were made by its most prominent members, discussing their political vision and tactics.

I have already touched upon the agenda of the Council of Ancestral Authorities, but despite its seemingly hard confrontational stance, its actual approach is that of both pressure and dialogue. In fact, Kalel emphasized during these meetings that, while being still in a frontal confrontation with the mining companies, the situation with the *alcalde* was more about making his words match his deeds. Their ultimate aim is to get recognition as a valid form of authority beyond both the communitarian and the party level. In fact, one of their current causes is to get a building completely dedicated to their reunions, something that the *alcalde* had promised (but not fulfilled), as Kalel said:

> We don't make electoral politics, on the contrary, you were appointed in an appeal to directly represent the interests of our communities. We don't make politics saying 'vote for me', we are not elected by virtue of any political propaganda. We want to be recognized by law, to promote unity. If the alcalde said: I have my functions, and you have yours. I would applaud this ... We need unity and development, to be together. We have a common cause for the people: this is what we want.

Beyond these tactical considerations, what is strongly shown here in more symbolic terms is how the Ancestral Authority council is somewhat of a replication, in the terms of Vogt (1965), of the traditional communitarian forms of authority, down to its ceremonial practices, but in a new context that is neither communal nor electoral. It is as if, in a way, modern activism had been encroached upon by the older conception of power, in a process that both transforms 'nativist' activism and traditional conceptions of power. But more on it will be said in the conclusion.

To finish this chapter, I would finally like to add some information about one of the most prominent and still active figures of the movement: Abraham Vicente, who evolved from a guerrilla fighter in the eighties to a member of the ancestral authorities today. It was Vicente who organized the commemoration of the referendum in 2018, and he is still a strong figurehead in local resistance politics. Some words on his political involvement with both the guerrilla and the new resistance are needed to understand how, in a way, the struggle against the military dictatorship has evolved into the struggle against neoliberalism and mining companies, which are supported by the military-supported right-wing political parties of today.

Abraham Vicente: From Guerrilla Fighter to Environmental Activist

Some days before the anniversary celebration, I received a call from Miguel Ángel. His daughter, a student at the Centro Universitario de Occidente (CUNOC), a centre of the University of San Carlos of Guatemala located in the near city of Xela, had to interview Freddy Rodríguez, agronomical engineer and professor at CUNOC, and native of Momostenango. Rodríguez was a victim of persecution during the last period of the civil war (1982–1996), and barely made it out of

Guatemala alive, becoming a refugee in Chiapas, Mexico. Miguel Ángel asked me to support his daughter with my recording and photographic equipment, but also had an interest in the material. So, we went to Xela to interview Rodríguez.

During the interview, Rodríguez talked about his personal life, and also about the activities of the guerrilla in Momostenango. At the time, many academics at CUNOC, especially those with Marxist inclinations or ties to the guerrilla movement, disappeared and were killed. He recalled how his family was intimidated by the armed military patrols, which prompted him to escape to neighbouring Chiapas. Luckily, he survived the army checks at roads and the border; in fact, at the time, carrying a copy of Marx's works was an offence punishable with death. After he escaped, the Mexican government was generous: they offered Freddy the opportunity to further his education at the National Autonomous University of Mexico in the capital and later relocated him as a teacher of agronomy in San Cristóbal de las Casas, Chiapas, until the armed conflict ended, a wait of fourteen years for him. Luckily, later his family made it to San Cristóbal, too.

During the interview, Freddy told us that Momostenango had an active guerilla movement, something that is still not well known nor studied, since most of the guerrilla activity and the corresponding armies' genocidal activity were carried out in the Ixcan and Quiché region.[14] Momostenango was considered a low risk or 'green zone' at the time, thus military repression was not based on genocide against the general population, as in other places, but on selective assassinations. Some people in Momostenango helped the guerrillas with resources and with their capacity, as inhabitants of a merchant town, of transporting resources and people: 'There was lots of repression. In the Quiché region full populations were razed. In Momostenango it was more about selective violence. For example, against *educadores populares* [propaganda commissars]. Lots of the people that disappeared were indeed guerrillas, some were tortured and denounced their partners, which prompted more repression.' Women from Momostenango were also commanders in the Frente Guerrillero, which operated in Xeqemeya, near Santa Lucía la Reforma. Many *educadores populares* were kidnapped and disappeared. However, the guerrilla movement was not as strong in Momostenango, in part because the dual agricultural/mercantile patrilineage system had been successful in fending off big plantations and other forms of capitalist exploitation, thus reducing the motivation for participating in the movement. The population, mostly Indigenous, was sympathetic to the cause, but also cautious, given the genocidal politics of the Guatemalan state. Thus, the guerrillas in Momostenango pursued the 'prolonged popular war' approach.[15] According to Freddy, Abraham Vicente was one of the commanders of the local guerrilla front; one of the other leaders is still a doctor in the community.

Some days after the interview with Freddy, I went to the antimining referendum commemoration, with Abraham Vicente at its helm. Curiously, it was celebrated right in front of the town's secret altar, in front of the town's main square. The charred facade of the place is a reminder of the ritual activity carried out there in secret by the highest-standing calendric priest, was a weirdly appropriate scenario. After a small parade through the town, there was a miniature version of the traditional 'deer dance' (*danza del venado*). While seemingly a gratuitous feature of identity and folklore, the subject of the dance is the relationship between the white owner of a hacienda and the *awajmundo* or spirit animals of the earth divinity, so it made an interesting contrast with the metaphorical subject of the struggle of the earth divinities themselves against mining companies. Abraham Vicente made an appeal in K'iche'. He insisted on how mining would bring an uneven and illusory development to the town: 'Wealth for a few years, misery after.' He finished his speech saying, 'Our ancestors have commanded us to protect this land and that is what we are doing'.

Afterwards, we went to a traditional meal with the Council of Ancestral Authorities, on the patio of a house in the town centre. During the meal, the members of the council discussed the wider national context. In those days, the presidential persecution of the anticorruption commission of the United Nations (Comisión Internacional contra la Impunidad en Guatemala, CICIG) broke out in different legal actions against Thelma Aldana, head of CICIG and former supreme justice. These actions would end in the unilateral dissolution of CICIG by the military-backed president, Jimmy Morales, a former TV comedian (cfr. BBC News 2019).[16] Thelma is well liked by the movement, for she was instrumental in the organization of the antimining referendum, meeting with the communities beforehand. Those opposed to the then-president Jimmy Morales had planned some demonstrations against him in Guatemala City. The members of the traditional authorities pledged their support. It was there that I realized that the political consciousness of the population was very strong, unlike the impressions of some anthropologists that worked in the region in former years, which emphasized apathy (Bocek 2006). On the contrary, political effervescence exists, but these kinds of discussion are usually not conducted before people not working in the relevant local organizations.

Seeing it and hearing Abraham's speech in K'iche' and Spanish was a strong reminder of the ties between current-day Indigenous resistance and the struggles of the eighties, and proof of the political metamorphoses of the speech of Indigenous resistance against the Guatemalan State and extractivism. While during the eighties Marxist literature dominated (Freddy told me it circulated through pamphlets and comics in Spanish, while propaganda in Maya was oral in nature), today the struggle has shifted towards a speech that defends Indigenous rights and their ties with both cosmologies and sacred practices. However, I would argue that, locally at least, this rediscovery or successful fusion of Indigenous identity with anticapitalist resistance was made much more powerful and effective when the more one-dimensional speech of the guerrilla, which was still too rooted in an identity-agnostic and universalist Marxism, was abandoned; the success of the referendum vis-à-vis the stagnation of the guerrilla in Momostenango is proof.

Until now, this chapter has remained close to its ethnographic narrative, in order to give a detailed and involved account of Miguel Ángel's political biography. However, now it becomes necessary to address some of the points that were outlined in the introduction, namely, to discuss the importance of this material regarding wider debates on Indigenous political practices related to nature, as well as anthropological debates about ontology, to try to devise a somewhat different position.

Closing Remarks

In contrast to past assessments, which attributed a sense of political indifference and passivity to K'iche' people (cfr. Bocek 2006), in this chapter I have described a political context of effervescence, engagement and resistance. In the introduction to this chapter, I have mentioned the recent ascendance of 'cosmopolitical' and ontological models in the study of Indigenous politics of resistance to extractivism, as well as some of the criticisms and risks inherent to these proposals. In general, these proposals have been extraordinarily successful at depicting the ontological differences of Indigenous and Western politics in regards to landscape and nature, but some of them have also somewhat tended to fall into common traps such as exoticization, ontological reification and a lack of grounding in real-life political processes. Thus, the first and most important precaution to be taken is to remember that political action in Indigenous communities is in no way predetermined by a reified Indigenous ontology, which would be all too similar to prior ideas regarding a reified holistic cosmology of nature.

Indeed, Miguel Ángel's career did not start out of an ideological commitment with a reified 'cosmology/ideology', or merely by being already within a predeterminate ontological outlook; instead, his political compromise started precisely in the same way in which engagements with cosmology have tended to appear throughout this book: in contexts of personal risk and social struggle. Miguel Ángel did not start his political career by defending ancestral Indigenous sites, he started his political career by an experience similar to that in which many others in town first relate to the cosmological: by recognizing himself as being 'inside' that cosmological dimension, a literal captive of the prisons of the earth beings or *mundos*, punished by such, which was the reason why he was sick and had bad luck in his life. Only by recognizing himself as included in that cosmological dimension, and then exerting actions that were not only social in nature but also cosmological, did he become a calendrical diviner, then a local authority and finally an activist. From then on, we can see Miguel Ángel constantly advancing in both 'worlds', guided by the strength of his own *nawal*, his *mundo* counterpart, which is also the force behind his political agency.

At a later point of our friendship, Miguel Ángel and I discussed the nature of his *nawal*. He told me that it did not have an animal form (unlike, for example, Rudy's eagle *nawal*), but it manifested in his dreams in a different way. For example, he told me that, as a kid and later in life, he would dream himself covered in the armour of the archangel Saint Michael. For him, that was the manifestation of his *nawal*. As readers who know Spanish will realize, the name Miguel Ángel is really an allusion to the Archangel Michael, so in a way it is almost a pun. Among *costumbristas*, Saint Michael is also regarded as a manifestation of lightning or *kak'ul ja'*. Miguel Ángel told me that he usually had dreams of flying with the armour of San Miguel, prefiguring the attempts on his life too:

> I entered the church in Momostenango and found soldiers and chopped their heads and hands with my sword. I dreamed that in the church, on the altar, there were demons, and they shot at me, and I fell. But I raised again. They then said among themselves: he is alive! And I escaped. This is what the *chuchqajaw* told me: this is your calling, this is your destiny.

Thus, it is hardly a surprise that Miguel Ángel's altar had multiple icons of Saint Michael brandishing his spear and covered in his armour. While Miguel Ángel perhaps did not share the same *nagualismo* ideas of other *costumbristas* like Rudy and Marcelino, he indeed visualized the force of his personality and destiny as tied to a *mundo* image, both his protector and his destiny (Figure 5.2.). In a similar way to how Miriam Lamrani suggested regarding the cult of folk saints in Mexico (Lamrani 2018), it was an image of his own inner life, externally manifested; however, it was also an image of his political, public life, as his assimilation of the

Figure 5.2. Miguel Ángel burns offerings at the patio of the church, before the image of Santiago, patron of the town. © Alonso Zamora Corona.

attempt on his life to the struggle of San Miguel against the demons. Thus, it is not only that the image is a symbol, but the image is somehow himself; as he said: 'I am *nawal* Kiej' (*yo soy nawal Kiej*).

Now, before developing this analysis further, one must consider Miguel Ángel's activism in relationship with other forms of power. One of the most common mistakes in some assessments regarding the ancestral authorities movement is the opinion that they are 'not truly traditionalist' or, even worse, that they represent merely a 'New Age' version of Indigenous religiosity.[17] However, what transpires in both Miguel Ángel's testimonies and the meetings of their members is that what these organizations do is to actually replicate the practices of traditional authorities and rearticulate them in new contexts, and, furthermore, they do it regardless of religious affiliation. Thus, there is hardly anything merely 'neo-Maya' about the members of the council, for many of them are indeed *costumbrista* members of the actual communal authority system and engage in traditional practices at the altars of their own. Instead, what is new is the emergent context of the political engagement into which they have taken the *costumbrista* cosmology, from a context of community organization described in the classical works of Maya ethnography quoted in the introduction to this book, to a confrontation with global extractivism.

In this regard, I have mentioned Evon Vogt's theory of 'structural replication' (Vogt 1965). Closely related to that idea was his theory of 'encapsulation', which he defined as 'the conceptual and structural incorporation of new elements into existing patterns of social and ritual behaviour' (Vogt 1969: 582). While seemingly describing a dynamic proper of a 'conservative' society, in fact, Vogt was not only thinking of it as a mechanism of native resistance to colonial religion and society, but also as some sort of dynamic of social change. As Vogt mentions, instead of confronting Catholicism or outright rejecting it, the Tzotzil Maya decided to 'encapsulate it' through their 'cultural' practices. For example, crosses were not rejected; instead, they became symbols of the ancestral caves where ancestors reside in the sacred mountains. As such, they were 'encapsulated' not only conceptually but formally, burdening them with symbols that allude to native cosmology, such as pine trees, flowers and the like, symbolic of the telluric divinities and their *nagual* abodes (cfr. Vogt 1969: 286–7). Similarly, Catholic saints were incorporated to the *nagual* cosmology, as protectors of the animal souls of people (as we have seen with San Antonio Mundo). Furthermore, this form of 'encirclement' of the new by the old was closely related to Vogt's idea of replication or repetition of structural patterns in many levels, since relationships of protection and patronage of the ancestors towards the people, shamans towards their clients, parents towards children, etc., were always characterized by the idea of 'embracing', which ultimately belongs to the 'corral' cosmology of *nagualismo* (1965, [1990] 1970: 125–33).

What is very interesting in the current day Momosteco case, and especially in the case of Indigenous activism, which is a new feature born out of the needs of the current geopolitical circumstances, is the fact that a certain kind of formal replication of the community authorities has been carried out by the Council of Ancestral Authorities. Ancestral authorities seemingly 'copied' the formal structure of communal authorities (*presidente, tesorero, secretario*, although using K'iche' terms: *kamal b'e, ajpwaq, ajtzib'*), as well as their rituals and symbols of power, by holding batons or *varas*, gathering in similar calendrical dates, burning offerings and making similar appeals to the ancestors in their gatherings. However, more than a nativist reappropriating of traditional authorities by activists, I see this more as if the traditional structures and cosmology of the communal level of power had performed an effective encroaching of our current day's 'geontological' conundrum, to quote Elizabeth Povinelli's conception (2016).

Thus, it is not merely the case that different cosmologies have clashed; the actual process is, from an internal perspective, that the Indigenous cosmological logic has somewhat engulfed or encroached our current geontological dilemma. Thus, my reading would be akin to that of Knut Rio and Annelin Eriksen's depiction of visionary leaders in Vanuatu, who have insisted that it is not only the leaders who have agency, but cosmology itself must be seen as something invested with agency in order to understand them (Rio and Eriksen 2014: 55). Thus, it is cosmology itself that acts here, expanding and entering in contact with new dimensions of social life and history. However, there is yet another factor to take into account, and that is the place of the individual and its empowerment by these movements, something that is often obviated in accounts of cosmopolitical resistance.

I have mentioned earlier how the previous and deceased members of a *cofradía* become like the dead ancestors that the new members of the *cofradía* must pray to in order to perform rituals successfully; thus, Rudy prays to the dead members of Sociedad Santiaguito to assure his success in future rituals. The word used to describe these deceased ancestors of the *cofradía* society is the same as the one used to refer to the gathering of the ancestors of the patrilineage: *may* 'the ranks' (of the dead). However, this idea is also starting to appear among the ancestral authorities movement, despite being of a relatively recent inception. During my stay in the town, one of the friends of Miguel Ángel, David Coxoj, died. He was the *nab'e kamalb'e* of the Council of Ancestral Authorities at the time, so his death was particularly felt by the council. He was well liked and respected because of his knowledge of the tradition. When Coxoj died, Miguel Ángel had a dream. He dreamed how a man dressed in white showed him a lightning in heaven. Then, a voice said the word: *may*. He was thus sure that his friend had become a *may*, a calendric ancestor. Hence, in a way, ultimately, this is how Indigenous cosmological power is made: to gain power, you must be favoured by the living, the earth divinities and the ancestors, in order to eventually join the latter and become part of the cosmological forces. It is especially compelling how this power is not only tied to traditional forms of government, which have somehow lagged behind the times. Instead, it has become ingrained in new forms of power. Thus, by dying while being a *nab'e kamal b'e*, Miguel Ángel's friend became a veritable *may* or ancestor of the organization itself, a cosmological power himself, just as a former *chuchqajaw* at a patrilineage would, or a *cofradía* president would: A sort of *dios mundo*, as Rudy said about his ancestors ('they are gods', *son dioses*). Thus, the final aim of the calendrical practice is not only to know about time in order to have power, but it is to become part of the order of time itself, in order to rule the earth.

Now, regarding the former in its relationship with recent engagements with Indigenous 'cosmopolitics', it is indeed important, but not complete, to only say that Indigenous people ascribe varying degrees of political and social agency to landscape features and thus they strive for their protection. This is true indeed, but it is not the whole story, at least not in Momostenango. The important thing is this: the agency of the sacred shrines and landscapes is part of a wider network that defines the political agency of Miguel Ángel himself. None is really subordinate to the other; in fact, as the case of Miguel Ángel's recently deceased friend shows, an activist is a prospective 'earth being' himself. Thus, it is not a nature that is merely an animated 'other' that Miguel Ángel is defending, but he is defending a nature that is the sum of his ancestors (being gods themselves) and their relationships with 'earth beings' or *dioses mundo*, the *e rajawal ri uwachulew* or 'rulers of the face of the world'. He defends a nature that is power, but also an earth that becomes *his* own power.

Therefore, we find the same dilemma that we have encountered in previous chapters. On one hand, we have a cosmology that 'acts' in a way, becoming determinant in social life, even expanding to new historical and political contexts. On the other, we have people acting in a cosmological fashion; that is, people who are invested with cosmological agency and, even more, who actively work towards obtaining it. The question is: who is doing which? As mentioned before, my conclusions will try to work out this final question, but for now it can suffice to say that not only daily endeavours or ritual endeavours in Momostenango are permeated by this paradox, but also political careers and activism. In the next chapter, this agency will be taken to a new level: that of Guatemala's national history.

Notes

1. For general overviews regarding contemporary extractivism in Guatemala and its relationship to social conflicts, see Morales 2014 and Véliz Estrada 2015.
2. Communal authorities are the local, traditional authorities of each hamlet. They are elected to serve for a year by a gathering of past holders (*pasados*) of the mayoral (*alcalde*) appointment. At the gathering, which usually takes place at the end of October on a Kiej day of the K'iche' calendar, two *alcaldes* are elected, each serving fifteen days. One of them is considered the main *alcalde* and the other is the vice-*alcalde*. The *pasados* also nominate people to fill the rest of the posts of the administration: *regidores* or assistants; secretaries who assist the *regidores* and keep notes of every meeting (*minuta*); a treasurer (*tesorero*) or bookkeeper; and many assistants to the *secretarios* in specific appointments, such as forest keepers, bailiffs and officials in charge of local services – water, roads, education and the like.

3. Municipal authorities are political authorities that are elected each four years in a democratic manner, through the nationwide party system. Once elected, the *alcalde* or mayor selects the members of the 'corporation', seven town councillors (*concejales*) and two *síndicos* or city managers. They meet each month with the eighteen communal *alcaldes*, the four ward *alcaldes*, and the local committees known as COCODE (Councils of Communal Development) and COMUDE (Councils of Municipal Development).
4. For example, the National Council of Indigenous Ancestral Authorities of Guatemala incorporates Xinca and Garifuna representatives, both non-Maya ethnicities (see National Council of Indigenous Ancestral Authorities of Guatemala 2009).
5. See Morales Sic 2004 for a general account of the 'Maya Spirituality' revival, concentrated in the idea of the 'invention' of tradition (Hobsbawm and Ranger 1983); see Bell 2012 for an ethnographic account of the movement among the Kaqchikel through the lenses of cultural studies, and Ponciano 2007 for the idea of Maya Spirituality as a 'religion'.
6. Health & Help, founded in 2005 by Viktoriya Valikova, which has a clinic in Chwinajtajuyub' (Sorokina 2017).
7. As explained in Appendix B, there are four days that can start a year in the K'iche' Calendar, Ik', Kiej, E and Noj; thus, they are called *alcaldes*, since they rule each year. The treasurer of the 'government of the Earth' (*gobierno mundo*) is Tz'ikin, thus it is a day related to wealth, while the secretary of the *mundo* is K'at, related to retribution.
8. This kind of action continues, but now under Mormon and Evangelic influence rather than Catholic one.
9. COIMO is an organization that coordinates Traditional Authorities to pursue specific development agendas, called Community Development Agendas. These are defined as 'instruments of negotiation' from the communities before the municipal government, as well as outside non-governmental organizations. For example, one such agenda is the Project of Food Sovereignty, which tackles nutritional problems (Serjus 2012).
10. In fact, the same word appears with the same exact meaning in De la Cadena's work, *Earth Beings*, despite it dealing with Runakuna or Quechua shamans (cfr. 2015: 94–95); it could indicate a certain convergence in the Spanish translation of Indigenous ritual terms in Spanish America, perhaps dating back to colonial times.
11. The expression could be rendered more literally as 'perhaps there is no order in this bundle (of words) that I present before you', ironically, a very elegant way to say that he is not being proper in his speech.
12. Robert Carmack obtained similar information. In fact, gold was mined in Buenabaj at some point (1995: 14).
13. Other projects were called Iralda II and 'Lucía'. Five concessions were granted to Reserva Caribe S.A, two to Pangea and one to Entre Mares de Guatemala S.A. (Chiquin 2013).
14. On the genocide of Indigenous groups by Guatemalan state in the eighties on a historical and human rights perspective, see Comisión para el Esclarecimiento Histórico 1999; for a collection of ethnographic accounts concentrated on the Indigenous perspective, see Carmack 1988; for a sociological perspective, see Vela Castaño 2014.
15. Regarding different guerrilla styles at the time, Fredy said: 'The FAR [Fuerzas Armadas Rebeldes, Rebel Armed Forces] followed a style of spectacular strikes that would demoralize the army and make them feel that they were not winning the war; other groups carried out tactics of protracted popular war, a war that would drag

on for many years, wearing the army out. Their philosophy was to minimize the suffering of the people, without exposing them to repressions. The EGP [Ejército Guerrillero de los Pobres, Guerrilla Army of the Poor] wanted the population to get involved. That is what caused the destruction of entire towns by the army, as it happened in Quiche, in Chu Pol, where they even dynamited the church … all men were killed.' Thus, the guerrilla approach in Momostenango was to minimize civilian involvement and causalities.

16. President of Guatemala from 2016 to 2020.
17. On some differences between *costumbre* and 'Maya Spirituality', ontologically conceived, see Kaputsa 2021.

6

Altars of History

War, Rituals and Social Memory

> Sorcerers and soldiers from Momostenango protected him (the president).
>
> —Luis Cardoza y Aragón, 'The Ineffable Tyranny'

> *El Señor Presidente* is not a fake story, it is not the fantasy of a novelist: He (the president) surrounded himself with Indigenous sorcerers, brought from the most famed places in the craft of magic.
>
> —Miguel Ángel Asturias, '*El Señor Presidente* as a Myth'

As mentioned in the introduction to this book, historically oriented chapters are fairly common in ethnographies of the Maya of Guatemala, past and present; similarly, the confrontation of ethnographic realities with change and history has been a fairly common feature of regional ethnography (Britnall 1979; Watanabe 1992; Watanabe and Fischer 2004; Wilson 1995). Similarly, it has been mentioned that Momostenango is perhaps the Guatemalan Indigenous community with the most detailed diachronic ethnohistorical monograph dedicated to it, Carmack's monumental *Rebels of Highland Guatemala* (1995); the K'iche' kingdom at Q'umarkaj, of which Momostenango was a part in the pre-Hispanic period, also has been the topic of an authoritative historical and reconstructive ethnography by the same author (1981). Thus, a historical overview of the town's past has been omitted in the present work; however, this does not mean that this ethnography will not engage with history at all, although it will do it in somewhat different terms to what is usually done in the regional context.

The topic of this chapter could be said to be what traditional ethnohistory called 'folk history' (Sturtevant 1966: 22–26) or 'the view a society has of its past' (Carmack 1972: 239), how people around the world portray, conceive and interpret past events. Of course, this opposition between 'folk history' and 'scientific history' has been questioned as being essentially Eurocentric since very early (Hudson 1966), leading to what Shepard Krech III called the shift from ethnohistory to 'ethno-ethnohistory', the 'attempt to fathom the structure of history and/or the historiographical conventions of Indigenous historians' (1991: 361), which implied, 'in essence, problematizing *historia res gestarum*' (1991: 361),[1] the traditional Western notion of history.

Following some of the classic works that follow these more questioning theoretical lines,[2] the placement of a historical chapter at the ending of this ethnography, rather than at its beginning, derives from a similar desire to explore the Indigenous idea of history, rather than present Indigenous communities as framed within the Western notion of history. Indeed, conceived withing the classical dichotomy of the historical vis-à-vis the 'socio-structural',[3] in most accounts of the past of Indigenous peoples, history has usually held the upper hand, despite the continuous defence that anthropologists have made of the 'structural' side of the dichotomy (cfr. Evans-Pritchard 1961; Lévi-Strauss 1966). In Western epistemology, 'structure' has been usually conceptualized as an effect of history and causality, while change has been considered superior to the 'primitive' denial of historicity that structure seems to imply, this superiority being considered as essential to the integrity of history itself (cfr. Rousseau 1975: 37–38). In consequence, when history usually appears in an ethnography, it plays the role of providing a background or some context to the understanding of the second: in general, thus, history has been conceived as a potential framing, a background for the theoretical arguments of anthropology (Bayly 2018: 109).

This chapter has a different orientation, deeply informed by the aforementioned critiques of orthodox historical postulates. In a similar way to Marshall Sahlins' reflection of the clash between Hawaiian cosmology and the expedition of James Cook (1985), this chapter will try to rethink the relationship between historicity and anthropology through the retelling of an event in which the cosmological and ritual practices of the Momostecos clashed with a historical 'outside'. Concretely, the subject of this chapter is the role of Momosteco soldiers and their sacred altars in the victory of Guatemala against El Salvador during the so-called Second War of Totoposte (1903–1906).[4] This story, which is still remembered among many people in Momostenango, is not only relevant today as a historical event of note, but also because the most important shrine in the town, the *awas rech tinamit*, a place that has crucial cosmological significance (cfr. Carmack 1998: 342–3; Tedlock 1992: 147), is said in the town to be connected with the death of the president of El Salvador at that time, Tomás Regalado, which took place during a skirmish with soldiers from Momostenango at Cerro El Sillón in 1906. To put it in plain (although not simple) terms, as many of my local informants told me multiple times that 'the altars won the war' and, even more, according to local lore, the ulterior destiny of Regalado is still connected to the shrine itself.

In order to assess the complex debate of non-Western historicity and its relationship with social life and cosmological ideas, in this chapter I will present first to the reader an account of the death of Tomás Regalado from the perspective of more traditional historical works, which are themselves fascinating when read in the context of what I have explained regarding cosmology in Momostenango since, as the reader will see, there are very concrete intersections between K'iche' rituality and the events narrated here. After this, I will explain how Momostecos today view the topic through different ethnographic testimonies engagements. Finally, I will to try to advance a proposal regarding cosmology and historicity in line with the one developed more generally in this book.

El Señor Presidente and *El Santo Mundo*

Manuel Estrada Cabrera (1857–1924), president of Guatemala from 1898 to 1920, has become one of the archetypes of the Latin American tyrant, thanks in no small part to the classic novel of the Guatemalan writer Miguel Ángel Asturias, *El Señor Presidente* (1946). His meteoric rise, from being a pauper on the streets of Quetzaltenango to the all-mighty ruler of a country, was almost supernatural for his contemporaries; his bloody rule, marked by megalomania, personality cult and repression, became in the masterful writing of Asturias the rich subject of a classic of Latin American 'magic realism' aesthetics.[5] However, there is a little-known and almost surreal episode in Estrada Cabrera's life that is only obliquely alluded to in Asturias' novel, despite the leanings of his writing towards the fantastic, and which is more deeply dealt with in one of its possible models, the non-fiction biographical book on the tyrant, *Ecce Pericles*, by Rafael Arévalo Martínez (1945). It seems that, according to the private accounts of General José María Letona and

Prince Wilhelm of Sweden, in the weeks before his regime was overthrown by protesters in April 1920, Estrada Cabrera consulted with Momosteco calendric priests regarding his destiny (Arévalo Martínez 1945: 560; Rendón 1988: 14; Wyld Ospina [1967] 1929: 185).[6] The question is: exactly when and how did the fate of Estrada Cabrera meet with the cosmological practices of Momostenango?

It all began in 1906.[7] Estrada Cabrera was down on his luck. Those who had fled Guatemala because of his repression gained at the time enough political and military support abroad to attempt a military invasion against him. Both Porfirio Díaz, dictator of Mexico, and Tomás Regalado, military *caudillo* of El Salvador, granted permission to the troops of the leaders of the rebels, the ex-president and general Lisandro Barillas and José León Castillo, to pass through their respective borders. But the first rebel invasion failed at Ocós (Rendón 1988: 184). Something more, in terms of firepower and political support, was needed to oust the tyrant. Pragmatic, cautious and a dictator himself, Porfirio Díaz did not commit to give any further support. But Regalado, romantic and ambitious, had grandiose plans. Having being president of El Salvador and being at the moment the most influential general under the government of President Pedro José Escalón, he dreamed of restoring the former Central American Union, formed in the nineteenth century by Guatemala, El Salvador, Honduras, Nicaragua and Costa Rica. By defeating Estrada Cabrera, he could not only retake power in El Salvador (he had already finished his democratic term), but in all of Central America. Escalón, a political realist, disapproved of Regalado's fanciful plan; in a display of bravado, Regalado bombarded the presidential palace of El Salvador with cannon rounds and departed to seek glory in Guatemala (Flores Escalante and Kuny Mena 2004: 191). But Estrada Cabrera had an ace up his sleeve: Momostenango.

Momostenango experienced a deep transformation at the close of the nineteenth century, a transformation that, while being led by a Ladino, in some respects resulted in a renaissance and expansion of *costumbre* and traditionalism. Teodoro Cifuentes (Figure 6.1.), a Ladino general with deep Momosteco roots, became the main political power broker in the community at that time (Carmack [1979] 2001: 303).[8] He is still known to this day as *El Reformador*, 'The Reformer', for he indeed reformed the town politically and even in its layout. He retraced the streets of the town, under his orders the new Catholic church was built and he built an imposing barracks building as the government centre of the town, which later became the seat of the municipal administration; even today, some people call still this building *el cuartel* or 'the barracks' instead of *la Muni* (short for *municipalidad* or 'city hall'). He also embellished and

Figure 6.1. Momosteco General Teodoro Cifuentes. Photograph by Emilio Eichenberger (circa 1909). Permission by Hansel de Paz, descendant of Cifuentes.

modernized the town centre, building parks and beginning the reconstruction of the town church. Thus, Cifuentes contributed towards a stronger centralisation of the life of the town. However, his work was not only military or administrative, as we will see, as his personal politics were also deeply related to the ritual life of the town. For, while a

Ladino, Cifuentes was deeply indebted to Indigenous elites, to the point that he used his political power to consolidate *costumbrismo* as the official religion of the town, and to consolidate its relationship to the local government institutions, as we will see.

As a general in the Guatemala army (Figure 6.2.), Cifuentes' power did not just derive from the local sphere, but also from his dealings with the national powers of the time, specifically from his personal friendship with Estrada Cabrera. His Momosteco troops slowly became a presence in the army of the time; according to Rendón: 'The Indian soldiers of Momostenango were reputedly the best in the country and were devoted to Cabrera' (1988: 193). This was not a coincidence, though. Estrada Cabrera, son of an Indigenous woman from Quetzaltenango (Rendón 1988: 18), was inclined towards meritocracy; in his propaganda, his rise from humble origins was a paramount example of social ascension. Hated by a sector of the Ladino elite of Guatemala City, Estrada Cabrera, himself the supreme leader of the Liberal party, presented in regards to Indigenous peoples in Guatemala a revitalized version of the Guatemalan *conservador* politics of the nineteenth century, which, while authoritarian, was also an interesting counterpart to the racial prejudices

Figure 6.2. Teodoro Cifuentes (fifth from the left) with his captains and some of his Momosteco troops. Photograph by Emilio Eichenberger. Permission by Hansel de Paz, descendant of Cifuentes.

of the progressives, who resented Indigenous people and considered them representatives of 'backwardness'.[9] To the contrary, the official ideology of Estrada Cabrera was deeply marked by the idea of merit and social ascension regardless of origins (which reflected his own personal destiny), which was combined with a fierce antidemocratic authoritarianism. Thus, an important element in his national ideal was 'the Indians that redeem themselves' (*los indios que se redimen*), the name given to the educated Indigenous elites that could both influence their communities and the state, the foremost example being the members of the Agricultural Institute of the Indigenous, *Instituto Agrícola de Indígenas* (Bertrán 1897). However, these local elites always had to follow closely the despotic dictates of Estrada Cabrera, who saw himself as an exceptional case of social ascension. For example, as a somewhat paradoxical example, Cifuentes (himself being a Ladino) was called 'an Indian general' by Estrada Cabrera, on account of both part of his origins and the men he commanded (Carmack 1979: 312).

On the contrary, Tomás Regalado was a representative of the *unionista* vision, which favoured a cosmopolitan Central American Union in place of localism; thus, he allied with Guatemalan émigrés in order to realize his vision. Regalado prepared to invade Guatemala in July 1906. After bombarding the presidential palace at San Salvador, he departed for Santa Ana, relatively near the border with Guatemala, where he gathered more supporters; afterwards, he and his troops reached the border near Chalchuapa. Regalado moved then into Guatemala: Metapán, Atescampa and Jérez soon fell to his army, and his troops approached Yupiltepeque. He intended to take Jutiapa, capital of the Guatemalan department of the same name, bordering El Salvador. However, Estrada Cabrera had previously ordered Cifuentes and Alarcón to move in with the regiments of Amatitlán and Momostenango to meet Regalado.

On the morning of 11 July, Regalado began to ascend the El Sillón hill, near Jutiapa, wanting to capture the Guatemalan flag there and raise that of the Central American Union. 'Follow me! Let's rally to the banner!' he claimed to his soldiers before calling a charge. But his luck was grim; a battalion of Momosteco soldiers was waiting for him in ambush. As soon as Regalado arrived holding the Union banner, a barrage of fire covered him. Seven bullets hit him, the first one in his face and the rest in the thorax. He died immediately. Even his golden watch was pierced by a 7mm calibre bullet; according to the legend, the hands mark eleven o'clock to this day (Flores Escalante and Kuny Mena 2004: 198). Oral tradition records that the name of the Momosteco who killed him with a bullet to the mouth was Tomás Chanchavac (FODIGUA 1991). Regalado's romantic folly had come to an end.

After the death of Regalado, El Salvador was forced to sign a peace agreement; Díaz, wily and diplomatic, even sent his congratulations to Estrada Cabrera. His victory collapsed the cause of the opposition in Guatemala for more than a decade; the Central American Republic dream of the *unionistas* was dead for good. *El señor presidente* would rule fourteen years more.

> As a result of the victory, the Momostecans were received as national heroes in Guatemala City and were further honoured by being selected as the elite guard for President Estrada Cabrera. Some five hundred Momostecan soldiers served in this capacity at La Palma, the presidential residence in Guatemala City, successfully defending Cabrera during many attempted coups until he finally fell in 1920. … During these duties in Guatemala City, Momostecan soldiers were given permission to construct outdoor pottery and stone shrines on the grounds of La Palma. (Tedlock 1992: 21)

In other words, they built *awas* shrines of the kind described in this and other ethnographies. In fact, four calendric priests from the town were elected as 'military priests' for the elite regiment, providing religious support for the soldiers by burning copal for them at certain special days of the K'iche' calendar (Carmack [1979] 2001: 312). Because of his closeness to native religion and practices, rumours started to surround Cabrera's success. As Cardoza y Aragón (1984) puts it: '*brujos y soldados de Momostenango lo guardaban*' ('sorcerers and soldiers from Momostenango protected him'). Thus, Estrada's Momosteco protection was rumoured to be not only military, but 'magical' in nature.

But *El Santo Mundo* would eventually abandon Estrada Cabrera. In 1917, an earthquake destroyed Guatemala City. The attempts at reconstruction failed, partly due to the corruption of his regime. Popular indignation grew. Resistance against him and the officialist Liberal party was in hands of a new political party, the Unionista party, incorporating both the Conservative opposition and disaffected Liberals. On 11 March 1920, a considerable pacific demonstration against Estrada Cabrera's government arose. The dictator committed then his last mistake: he ordered his army to fire against the demonstrators, rousing popular indignation. According to his biographer, Arévalo Martínez, that night, during his plea, '(Estrada Cabrera) locked in with some Momosteco sorcerers to divine for him his fortune in the war luck with the *unionistas*' (Arévalo Martínez 1945: 451). According to the testimony of his close friend, General José María Letona, 'He spent hours in the chapel at La Palma, kneeling before catholic images, praying with fervour with his head between his hands. He then consulted with the Indian witches who he ordered to come from Momostenango and Totonicapán, and locked with them to practice sorcery' (Wyld Ospina 1929: 185). But the results of the divination were negative. Protective rituals were in order: 'All night, in the piece where the shamans of Momostenango were locked with the most excellent *señor licenciado* and the most excellent *señor presidente*, whispers were heard, lamentations, invocations in the Indigenous language … a persistent smell of incense came out from the locks of the doors … Don Manuel felt lost … the demonstration of the 11th had scared him' (Hübner 1992: 459).

Estrada Cabrera's regime did not last long after this; through violent repression, he precipitated his own downfall. The National Assembly of Guatemala declared the tyrant unfit to rule. Estrada Cabrera resisted, and fought from his residence at La Palma, but by that time he had lost the support of key sectors of the army. He surrendered on 14 April 1920. Before the downfall, Cifuentes had betrayed Estrada Cabrera too, choosing to back the *unionistas* against him. Nonetheless, the Momosteco soldiers hesitated to follow his orders, for they knew that Ladino tyranny would intensify with the *unionista* victory; in the end, though, they abided (Carmack 1979: 297). One of the defining moments of Estrada's loss of power occurred when he was moved to a military academy after being deposed. There, the Momostecos saluted him with their weapons as usual when he arrived, but they lowered their rifles after some orders from Rogelio Flores, a moment dramatically narrated by Arévalo Martínez.[10] Estrada Cabrera had lost all his remaining power at that very moment. But he did not flee; he faced prosecution, still convinced of his righteousness. He remained under house arrest at La Palma during the rest of his life, while preparing his legal defence, finally dying from pneumonia in 1924. Momostenango's military achievements remained recognised, though, and its soldiers retained the privilege of being elite presidential guards during the government of the following dictator, Jorge Ubico (1931–1944).

The memory of some of these events is still alive in Momostenango. Certainly, the historical influence of Teodoro Cifuentes is still present there, and is widely debated. His descendants have pride in his military achievements, but dissident voices say that his rule was a bad time for Indigenous people; forced labour was used to build the barracks and the new church, Momosteco soldiers were badly paid to the point of being exploited and military discipline was harsh, with corporal punishment being common (Historia de Momostenango Blog 2008). Meanwhile, Ladino oppression against Indigenous people was rampant during the period, which Carmack called a 'despotic government' ([1979] 2001: 306; 293). Thus, while they earned a newfound national respect, the harshness of the conditions that Momosteco soldiers endured was the reason why they did not concern themselves much with Estrada Cabrera's fate during his downfall.[11] Therefore, for Momostecos today, the key event is the victory against El Salvador. And for them, while brave, their soldiers did not win that war alone. Once, while discussing the story of the war, my friend Abraham told me that one of his friends, a man who lives near the Pak'lom hill (near the town centre) and served in the army, told him some of the lore surrounding the Salvadorean war:

> His grandparents told him that, during the war between Momostenango and El Salvador … not only soldiers fought, but also calendric priests in the burning altars did, the people claim that they had power, that they had a great significance, that during that war people made offerings because Momostenango is the place of those who burn offerings … most of the people that live here did, they helped the soldiers, they protected them and the war was won, for most of us are Indigenous, that is what my friend told me … those who burned offerings fought too, for it was thanks to the burning-place shrines that the war was won.

The testimony refers not only to the personal altars of the people, but especially to the ones built by Cifuentes near his newly founded barracks, among which there is one particular altar to which this victory is attributed. It is not only the focal point of this story, but also corresponds to the summit of the ritual hierarchy of the town to this day: the *awas rech tinamit*, the main calendrical altar at Momostenango. Its lore, which transports us from the plane of the Western vision of history to the Indigenous one, can also mark the transition in this chapter from 'history' to 'structure', from history to ethnography.

An Altar and Its Lore: The *Awas Rech Tinamit* or 'Town Shrine'

As explained in earlier chapters, the shrines of the Momostecos have a hierarchy that mirrors the political hierarchy of the town. At its base are the patrilineage altars, then come the altars at local *auxiliaturas*, and at the top of this hierarchy is the *awas rech tinamit*, the town shrine. As in a domestic unit, there are two of them: a *warab'al ja* for the ancestors of the community (dead diviners and political leaders), and a *meb'il* for the increase of wealth. These altars are tended by the *chuchqajaw rech tinamit*, the highest-ranking calendric priest of the community; he is officially the calendric priest of the town *auxiliatura* or Indigenous government as well as the *corporación municipal* or the town mayor's corporation, and thus is concerned with protecting both the whole town and the town political officers. As was the case with the altars of the patrilineages, some of the deeper meanings of this place, both in lore and in reality, have eluded the enquiry of past ethnographers; my own approach to the subject was gradual, being somewhat covered by a net of rumours and hearsay, and thus it is worth recalling these circumstances in an ethnographic fashion.

One day near the start of my fieldwork I was drinking a beer at the tavern of Doña Bertha, an older, green-eyed lady of Ladino origins who owns a local tavern in a street to the left of the main church of Momostenango, a block or so away from the entrance to the main market of the town. She knows quite a bit about the town; she witnessed the development of Momosteco tradition from the middle of the twentieth century. She still remembers Robert Carmack, too. She is well liked and respected in the town. She likewise respects Indigenous traditions, although she personally aligns with neo-Catholicism, rejecting Indigenous elements like the Indigenous patron Santiago. During the patronal *feria* I found her, for example, following the non-*costumbrista* procession to James the Great. Her tavern is favoured as a place to drink a beer by many of the Momosteco workers before they head to their homes in the northern mountains via the numerous pick-up trucks outside.

It was in that unlikely place that I first heard the strange rumour that persists in town regarding the *quemadero de la Muni* ('burning place of the town hall'), one of the informal names of the town's altar or *awas rech tinamit*. Some further historical context regarding this piece of lore is needed. The remains of Tomás Regalado, currently in the general cemetery of Santa Isabel, are a subject of certain controversy to this day. It seems that Regalado's body was badly damaged by the bullet barrage that killed him; furthermore, it was kept for a whole month in the possession of the Guatemalan army at the tropical fortress of El Jícaro, although it seems it was embalmed. Popular rumours, which persist today, affirm that the Guatemalans did not give back his actual body to El Salvador. Historians consider these rumours to be unfounded, since the body was recognized by international observers at the time; however, DNA analysis has been forbidden by the family (Marroquín Gálvez 2014). In any case, for the Momostecos, as Bertha explained to me, this suspicion is replaced by an unsettling certainty. When I asked her about the town's shrine, Bertha gave me a strange reply involving Tomás Regalado himself: 'They cut his head off. They gave a false body to the Salvadorians. Then, they buried the head here in Momostenango. There, in the altar of the town hall, under the burning-place for offerings. Because the head is the oracle of the Indians.'

Many versions of this story run within the town, not only among Ladinos. I gathered another one in K'iche', from the mouth of a direct descendant of Teodoro Cifuentes, who, despite being related to this powerful character, leads a modest life as a merchant. He frequently heard the stories about his great-grandfather's military prowess, and even mentioned that there is a 'Blue Book' with some of the oral tradition of Momostenango, where this episode is consigned:

> My great-grandfather is the general Teodoro Cifuentes, he was a landowner and owned many houses. But I have no land, for he had many women, and they squabbled among themselves, and thus my grandfather sold his land ... My great-grandfather had military comrades, and when he fought alongside them, they took the head of the captain of the army of El Salvador. My great-grandfather cut his head and they brought it to town: [they said] here it is, the head of our enemy! And the head is buried in the commerce locals [in front of the main plaza of Momostenango]. The place is always closed because the head is buried there, where the altar of the town is. Tomás Regalado was the name of the leader of the army of El Salvador. This story is in the 'Blue Book', many stories about the town are there, but I don't know where is it and where the book is, there are lots of stories about Tomás Regalado there.

I was able to gather a bit more on the lore regarding this episode during my stay in the town. According to another Indigenous informant, a well-read and knowledgeable middle-class man living in the town centre, there are two altars, one at the commercial establishments in front of the main plaza of the town, and another one inside the municipal corporation building, which was originally the barracks of Cifuentes. It seems that the altar at the town hall served originally as a war altar of sorts. The calendric priests would perform war divination for Cifuentes there, blessing his troops through calendric rituals and, even more, they would elect young people for forced conscription through calendrical divination. He repeated the same information told to me by Miguel Ángel regarding the sacred days that were used for war magic in Momostenango, but added something else:

> They burned for the soldiers in Tijax and Imox. Tijax to bless their weapons. They blessed the rifles presenting them before the *quemadero*. Imox to make them invisible. And they burned in Ajpu, the blowgunner, the hunter of *El Santo Mundo* [*el cazador del Santo Mundo*]. For the hunter guided them in battle, but if others had burned in his behalf, he could have betrayed them and led them astray.[12]

He added that it was the power of these rituals that allowed the Momostecos to 'become invisible' and undetected during the ambush that ended Regalado's life. I asked him about the idea of Regalado's head being there. He knew about it too. 'It is the oracle of the mountain', he said. 'When there is an omen for the town, the head talks to the guide of the people, the *chuchqajaw rech tinamit.*' He then pointed at the door where the altar is, clearly visible from the room where we were sitting at the town hall during our interview. 'See that charred door over there? That is where the town guide burns offerings. It is never opened to the public.'

Thus, for many Momostecos the victory over El Salvador was a miracle of *El Santo Mundo*, but it is also a miracle associated with the reformation of the town, and with its history. It is eerily fitting that Regalado's remains became (or are rumoured to have become) a trophy/vessel for that divinity. Furthermore, Cifuentes reorganized and reinforced the role of *costumbrismo* in the town, for it helped him to further his military vision. But there is another, more mythical background that is important to mention for the sake of our argument. According to Robert Carmack, at the town's main altar, powerful historical figures are adored as earth gods or *mundos* themselves: 'Important political figures of the past (such as cacique Diego Vicente and presidents Rafael Carrera and Jorge Ubico) … are propitiated by the town priests [there].' (1998: 343). I myself heard on one occasion that the spirit of Tekum Umam, the last K'iche' king defeated by the Spanish conquistador Pedro de Alvarado, was there. In fact, the group of these characters have a name: they are called *los principales del Mundo*, the Indigenous term being *ajawib' rech ri Mundo*, 'the lords of the Earth'. According to Garrett Cook, they are a mixture of military men and characters from the *Danza de la Conquista* (Conquest Dance), which commemorates the defeat of Tekum Umam and his noblemen and diviners (*ajitz*) before the Spanish:

> The *principales del Mundo* consist of a number of military men: captains, colonels, sergeants, corporals, sublieutenants and a band of soldiers who, all together, are the guards of *Dios Mundo*, and behind these soldiers come *tzitzimites* (diviners), the Rey Quiché and Tekum, all of whom are characters in the Conquest Dance. Tekum is said to be the Mundo, and it is he who speaks for the group. It should be pointed out that the narrator of this tale, and in fact most of the powerful Indigenous families from Momostenango proper, that is, the cabecera, have family histories of militarism. Thus, these soldiers of *Dios Mundo* have very strong prototypical connotations in Momostecan culture and may perhaps be ancestral figures. (Cook 1981: 377–8)

In fact, as Cook clarified, he gathered this information from a myth recalled by a man whose father was a captain, presumably in the war against El Salvador and who performed *costumbre* himself, 'to defend the soldiers' (1981: 654). Furthermore, he mentions that, in fact, the good fortune and survival of the Momostecos in war is related to the mythical origins of *costumbre*. According to him, in the primordial time people started disappearing; they were being eaten by the animals of *El Mundo*. In their plight, people find the *kaq k'oxol*, a small red god with an axe who is the patron of all diviners. He taught *costumbre* to the first town elders, and explained the calendar. Then, he introduced them to the *principales del Mundo*, who created the appointment of the town priest or *chuchqajaw rech tinamit*. They also created the barrios or *calpules*, and bestowed upon the elders the two main images of the town: San Antonio Mundo and Santiago. Furthermore, this author suspects that the myth gathered by him is closely related to the influence of Cifuentes in the town:

> As patron saint, Santiago is for some Momostecans the paramount deity, and his fiesta is today the most lavish public ritual. His prominence though seems recent, and while now supported by Momostecan tradition was possibly the result of what might today be called Ladinoization in the time of General Cifuentes, that is, the early twentieth century. When Cifuentes ruled Momostenango as caudillo, Santiago became a patron saint in the Spanish Catholic sense. The *patrón* is seen by Momostecans as a cacique and a war captain, and Momostenango has a history of domination by allied Ladino and local Indigenous militarists. In pre-Cifuentes days the *patrón* appears to have a *parcialidad* image maintained on behalf of the *pueblo*, rather than the property of the *pueblo* maintained by a *cofradia*. (Cook 1981: 400)

Thus, we have a clearer image now. It seems that the whole office of the *chuchqajaw rech tinamit* or town priest is tied to a history of militarism and its corresponding mythology, being created by the *guardia del Mundo*, the Indigenous lords of the court of the last K'iche' king, at the helm of which is the *rey del Mundo* ('king of the Earth'), as Tekum Umam is also called in the myth (Cook 1981: 664). But there is ambiguity in this: as Garrett Cook recounts, the raconteur vacillates on whether to include Pedro de Alvarado himself there, and probably that is the reason why Guatemalan presidents sympathetic to the town, such as Rafael Carrera and Jorge Ubico, are reported to be worshipped at the town altar. With this context in mind, we can turn to the next ethnographic question: what is the actual significance of this place for the present life of the town?

The Cosmology of the Town Shrine

In early 2018, the past *chuchqajaw rech tinamit* of Momostenango, José Itzep Herrera, passed away. A couple of months after the death of Itzep, the election of a new town calendric priest was arranged. The event took place at the Municipal Salon of Momostenango, a place for political gatherings (it was there that the main reunions for the antimining referendum took place, for example) on 12 March 2018, or 13 Kiej in the Maya Calendar. The traditional election, which implies a long session of divination in which four calendrical priests participated, came out in favour of Miguel Vicente, a descendant of the founder of the town, the seventeenth-century cacique Diego Vicente, whose family are in a way the last representatives of the pre-Hispanic Indigenous nobility (Carmack [1979] 2001: 310).

A while after his election, in the interregnum period between his appointment and his investment, I managed to arrange an interview with Vicente, thanks to my friend Miguel Ángel Pelicó, whose career was the subject of the former chapter. As a confirmation of his prestige and his authoritative character, mentioned in the earlier chapter, a mere phone call from him sufficed to arrange the interview, something that impressed me. The interview, conducted in K'iche' with the help of Miguel Ángel, at Vicente's home, was long and conferred a wealth of detail regarding rituals and dates associated with Vicente's future appointment. Of course, the readers of this chapter would wonder, as I did, what did he have to say about the Tomás Regalado story? The answer was somewhat disappointing: Vicente neither confirmed nor denied the story, although he indeed mentioned that Regalado's head was cut off and taken by his Momosteco captors. My translator was of the opinion that Vicente politely changed the subject, and I did not press on the matter, so, sadly, I could obtain no confirmation for this part of the lore from Vicente himself.

However, Vicente offered a crucial piece of information regarding the cosmology associated with this altar, with many implications for this book. The town shrine, just as the altars of the patrilineages, is itself a 'corral' where the soul-essences (*nawales*) of all the inhabitants of the town are contained. Thus, the main function of this altar is to protect the town and communities from the actions of outsiders with evil intentions or 'people that come to cause harm', which are identified with wild animals like coyotes and snakes, or *awaj mundo* spirits. The thirty-two communities that form Momostenango must pay a contribution (*pajo'm*) to the town altar or face being 'expulsed' from the altar by being omitted from the prayers of the town calendric priest, something that will bring them trouble and harm.

> The town shrine is like a corral, all people are there, all Momostenango is inside it, and for those who remain outside, it is as if a coyote arrives and eat them, they start having problems ... The thirty-two cantons are invited to give their contribution, their help.[13] These are those whose names are

> there [inside the altar], that is the reason why they are inside the corral, they are inside, not outside, and those who won't help, are outside of it, for they won't help the earth gods, the earth and the ancestors.

Vicente made sure to repeat his explanation in Spanish and left little doubt that the corral cosmology that exists at the base of the social structure is clearly 'replicated' (in Vogtian terms) at the top of it, for he would use the expression *corral del pueblo* (town's corral) to refer to the shrine itself. Furthermore, Vicente said that the 'corral' can be understood to be a 'circle' (*sipaniq*) or even a 'box' or place for safekeeping souls, for it is nothing more than 'our house' (*qachoch*), the cosmic house where the animal souls of the whole community live.[14] Thus, cosmology seen from the perspective of this altar is like a ring of concentric circles, where shaman-priests at the bottom of the hierarchy protect their own families, those who are associated with particular towns do the same for their local *alcaldes* and communities, and the town priest does the same for both the mayor of the town, the political hierarchy and the thirty-two communities that form the town. In this fashion, the answer to the problem of whether the hierarchy of shaman-priests is simply political, as Carmack considered (1998: 342), or ritual, as was the opinion of Tedlock (1992: 35), is that it is actually both. The ranking of the Momosteco shamans is defined by the socio-political hierarchies that they are associated to, going from the lowest level (the patrilineage) to the highest (the town's government), but it is ritual too, since the highest-ranking altar 'contains' the ones that are beneath it and protects them in a spiritual way.

In this regard, Rudy told me, when asked about the town priest, that 'the people don't know, but he binds us all, everyone is bound by him'. At the time, I did not really understand much about cosmology in Momostenango, so I did not know what this 'binding' referred to. But after talking with Vicente it became clear: the town's calendric priest is just like a patrilineage priest, but he has a direct link with the ancestors of the whole town (the *oxlajuj may*) as well as dead political figures of the past, the *principales del mundo* mentioned before (the last K'iche' king, Tekum Umam, the K'iche' elders of the past and Guatemalan presidents), and he can decide which community is 'inside' and 'outside' the 'corral', leaving their spirit essences open to the attacks of the forces of the underworld (incarnated by the foreign enemies of Momostenango) by 'writing' or 'erasing' their names from the altar. As the reader can see, this protection is ambiguous; it also duplicates the protection offered by the *principales del Mundo* to the first human elders, as described in the myth narrated by Cook (cfr. 1981: 656–7; 674). In fact, one of the first things I heard when I arrived in the town regarding the altar at 'La Muni' was that there were two types of offerings made at the place: the first, asking for rains, so the rain will be gentle and not kill people nor destroy crops; the other offerings asked for political order, 'so the people will not raise against the government'.

Thus, the 'binding' action of the town priest, mentioned before, is cosmological too, for the cosmology of the town is like a series of *corrales* or sacred underworld spaces where the animal co-essences of the people are held, much like in Evon Vogt's account of Zinacantan, discussed earlier (Vogt 1969: 384). However, instead of a single sacred mountain that controls these essences, as in Zinacantan, it seems that in Momostenango this dynamic is 'replicated' from the bottom to the top. This would constitute an even stronger example of the principle of 'structural replication' that Vogt proposed for the Zinacantecos and the Maya in general (1965), since this replication is perfect in each level and is both political and ritual, not only ritual, as he described in Zinacantan. Furthermore, curious is the fact that the word *pet* 'to embrace or encircle' is used by the Tzotzil to describe these structures as Vogt described (1990 [1970]: 130), while the K'iche' use the word 'circle' (*sipanik*), and the action of being kept inside them as 'binding' or 'wrapping' (*keb'otzik*). Moreover, the Tzeltal describe the inner space where the 'bird of the soul' is within the human heart as a 'box' (Figuerola Pujol 2010: 43), in the same way that the K'iche' describe the corral space within their altars; thus, the micro-cosmical space of the body/soul is replicated in the macro-cosmical space of the social.

Discussion

This chapter has read until now like a disparate collage: the almost surreal national history of Guatemala, which inspired Latin American *realismo mágico* literature and reads like an example of it; the local lore of Momostenango, and the deep cosmology behind it. While this panorama seems disperse, it can be said that the antinomies mentioned at the beginning of this chapter remain at the core of all these testimonies: history and anthropology, diachrony and synchrony, past and present, Western versus Indigenous visions of history. Thus, examining this event could lead to a deeper understanding, in the context of Guatemala, of how Indigenous people see historical events in regards to their cosmology.

In the story of the death of Tomás Regalado, we can sense some analogies with the Sahlinsian retelling of the death of Captain Cook, but the analogy breaks in other senses. While Sahlins' analysis is deeply informed by present and past ethnographies of Hawaiian societies, he could not directly interrogate anyone on how Captain Cook's death was integrated into native political/ritual structures and oral memory; he had to work with historical testimonies alone. In my case, the problem is perhaps that history becomes at some point lore and legend: Regalado's story, while true to the point of his death, becomes apocryphal when dealing with him becoming a war trophy, supposedly buried in the main altar of Momostenango, a fact that I could neither prove nor disprove. Thus, in Sahlins' example, Captain Cook's death is historical, but the 'interpretation' or 'symbolism' is contested by his critics (Obeyesekere 1992); while in mine, the 'symbolisms' are clearly articulated by the informants, while the 'facts' are, perhaps, doubtful.

However, beyond the level of historical 'facts', the main problem is theoretical, and this is where the material presented, in terms of both the historical and the 'structural' side of the dichotomy, could make a difference. In Sahlins, culture is used to break the antinomy of structure and history, positing a world of culturally determined 'categories in action' where human experience is refracted and appropriated by 'cultural concepts' (1985: 145). Captain Cook's death is posited by Sahlins as an example of cultural categories informing 'symbolic action', the latter conceived as a means of escaping the antinomies of history and society by achieving a new synthesis. The lasting contribution of Sahlins' work was to show that, at some point, the dichotomies posited by classic structural-functionalism break: 'what is functional, in this sense of instrumental, must be structural' (1999: 408), an assessment with which I concur. However, culture remained a somewhat reductionist anchor for his new synthesis: 'Desires depend on historical contexts of values, on existing or potential relationships of the culture, not only for their content but for their possible realizations' (1999: 408); thus, for Sahlins, history was determined by symbolic actions informed by culture, encounters like those between the Hawaiians and Cook being instances of 'a highly dynamic form of cultural scripting, dialectically active as a moment of conjuncture between existing terms of reference and the novelty of what is experienced at the moment when those perceptions are deployed' (Bayly 2018: 119). Thus, the Sahlinsian synthesis was that of history and structure within culture.

Building upon the debates generated by the work of Sahlins, Marilyn Strathern's proposal ([1990] 2013) also resonates strongly here. Indeed, in her reading on the encounter of Melanesians with European colonialists, heavily influenced by Roy Wagner's *The Invention of Culture* (1975), Strathern has also sought to overcome the dichotomy discussed in this chapter by redeploying the debate along a different line, that of artifacts. Indeed, Strathern has proposed to consider events as *artifacts*, something that is somewhat surprising, but that resonates particularly with this case. According to Strathern, Western accounts of Indigenous material culture are usually based on contextualization, which is a similar procedure to that which explains 'events through structure' and historical events through contexts. In contrast, Melanesian images do not really work within the Western mould of context and object, but more in a relational one, where an artifact is an occasion of both exchange relationship and the creation of more images. Thus, Strathern argues that 'we should extend our concept of artifact to performance and to event' ([1990] 2013: 174).

Obviously, Strathern's idea resonates powerfully here. Indeed, in a way, in Momostenango, history *is* altars; it is not only that the altars are vessels of the historical, but, in fact, history itself is produced by the altars, just as cosmological actions. The Momosteco built altars to win the war, and then the war itself changed the arrangement of the town's cosmology. Thus, events here have an artefactual dimension: history could be said to be an artifact whose use is to create more altars, more images, refracting into new times and new people. As I have mentioned, the most important thing to have in mind is that, for the people of Momostenango, it was not soldiers alone who won the war, but the shrines or altars. Of course, this means that the ritual, and therefore the cosmological action of the collective, had a decisive influence in the outcome of the war. Having achieved this victory, the outcome of the war itself changed the ritual organization of the town, impacting its 'structure': the shrine of war became the main shrine of the town, and the place where the head of the vanquished enemy lies became the main shrine in Momostenango. Thus, if we were to analyse this, we would have a sort of loop, as those described in other chapters of this work: structure impacts history, history impacts structure. However, this perspective is still too Western. The truth is even more complex. History is not dissolved nor merely impacted by structure. In the Indigenous perspective, history is *caused* by the agency of structure, and not the other way around. Thus, history is a sort of 'sorcery', nothing but the magic of structure, the magic of collective ritual action.

Thus, Strathern's intuition starts to resonate powerfully from this point, for we can see the affinities and difference with the Western idea of history. As Strathern states, Western history is 'context', it seeks to dissolve individuals and their deeds in the flux of structures and rational explanations. The Indigenous idea, while akin to it, is more radical: it dissolves events themselves into the flow of what we may call 'structural magic', but it takes this a step beyond, because, and unlike Western structures, which are merely contexts of backgrounds of historical action, Indigenous 'structures' behave like agents themselves and become *artifacts*, concrete beings.[15] In conclusion, we could say that the idea that 'everything is written' in contemporary Maya altars (cfr. Tedlock 1992: 80) that has been found once and again in various chapters of this ethnography, must be understood in a historical way too, and in an active way. Things are not merely written on artifacts; history is written *by* the agency of artifacts too. And herein lies the unending sense of surprise and the sense of invention that Strathern talked about when referring to the apparition of new artifacts within Indigenous life: 'cargo cults', ritualized transactions, or dances, to name a few, all of which contain history, but are also history and even historical agents themselves ([1990] 2013: 170).

Notes

1. 'Historians are in the habit of distinguishing between *res gestae* ["things done" in Latin] and *historia rerum gestarum* ["the history of things done"]. The former is the totality of everything that ever happened; the latter is the narrative of events that happened' (Muntz 1956: 2).
2. Some classical works within this current would be Fogelson's studies on ideas of history among the Cherokee and other Native American groups (1974, 1984), Rosaldo's work on Ilongot head-hunting (1980), and, of course, Sahlins' *Islands of History* (1985), which will be of great importance to this chapter.
3. The dichotomies that have characterized this debate have been diversely named: 'structure and process' (Vogt 1960), 'hot societies' vs 'cold societies' (Levi-Strauss 1966: 233–4), 'events vs. non-events' (Fogelson 1989), and, of course, the one this work follows the most, 'structure' and 'history' (cfr. Sahlins 1985: 136–56).
4. Spanish for 'toasted corn tortilla', from the Nahuatl word *totopochtli*, 'toasted thing'. This food was the staple food of both armies during the conflict, and the scarcity of corn that the war caused led to it being called thus, somewhat scornfully, by the population (Rendón 1988: 183).
5. Perhaps the most important current within Latin American narrative of the twentieth century, which concentrates in the category of the 'strange' rather than the veridical or the purely fantastical (cfr. Anderson Imbert 1972: 2), and to which the work of Miguel Ángel Asturias could be said to pertain (Couffon 1962).
6. Asturias, who makes a fictionalized and not a historical account, only referred to this episode in his novel with a passing remark about 'the pagan Indigenous witches that dance at his home [the president's]' (Asturias [1946] 2000: 310).
7. This account of the Second War of Totoposte is based mainly on Rendón 1988: 183–96, as well as Flores Escalante and Kuny Mena 2004.
8. Following the procedure of using pseudonyms for key characters and informants, Carmack refers to Cifuentes as 'Cienfuegos' in his book on K'iche' social history (1979).

9. Arevalo Martínez's book is an example of the racial prejudices of the *unionistas*, as recent studies have shown (Aburto Delgado 2019). One of the accusations made against Cabrera and his nineteenth-century predecessor, the dictator José Rafael Carrera Turcios, was that of the 'scandal' of being ruled by a non-white *caudillo* (Woodward 1993). In the case of Carrera Turcios, he is even acknowledged as 'having taken the interests of Guatemala's Indian majority to heart' (Adas, Stearns and Schwarz 2009: 67). Estrada Cabrera was more of a personality-cult inclined authoritarian with less redeeming features, despite his meritocratic inclinations; nonetheless, Ladinos deeply resented his authority on the basis of his non-whiteness.
10. 'Having arrived before the two ranks of Momostecos, general Enrique Arís ordered: For *El Presidente*, present your weapons!
    Do not! – Interrupted Rogelio Flores – Here goes no president of us: Just a prisoner of the people' (Arévalo Martínez 1945: 588).
11. Ubico, equally despotic, had a better reputation; he was called *tatanol* (great father') and was considered a 'protector' of the Indigenous people (Carmack [1979] 2001: 306).
12. This testimony is interesting because it links a more native divinity to the war; Tedlock had only mentioned that Santiago, the patron of the town, had lent magical protection to the soldiers (Tedlock 1992: 21).
13. The word *pajo'm* comes from *pajik*, 'to weigh', and in this context alludes to the benefices given to the town calendrical priest in order to ensure he can do his job (mainly money and ritual materials like incense and candles); however, as Carmack mentions, regarding their own living expenses these priests are wholly self-sufficient (Carmack 1998: 341).
14. Of course, these conceptions are completely comparable to those found by Evon Vogt and his team among the Tzotzil, whose main shrine at the sacred mountain of Bankilal Muk'ta Vitz ('Older Brother Large Mountain') is similarly conceived as a house where the ancestral gods keep the animal co-essences of the whole town (1969: 384–95).
15. For example, the epigraphist David Stuart has stated how historical events are not the true focus of the inscriptions of the Classic Maya era: things are. Indeed, Maya inscriptions were not made in order to commemorate events, but the creation of artifacts, and their consecration, the lives of kings being merely preambles to the awakening of such objects to a life of their own: 'A concept of history where objects, through ritual associations, often came to rival personages as focal points. Maya history is clearly in many ways a record of events and people, but it is perhaps more accurate to see it as a "history of things"' (Stuart 1995: 154–5).

Conclusion

WHAT IS COSMOLOGY?

Rather than presenting a general outline of K'iche' cosmology in a traditional fashion, this book has presented an ethnographic collage of Indigenous cosmological engagements from the personal to the political and the historical. However, now that its ethnographic narrative is over, it is necessary to ask ourselves if a more general sense of what is a cosmology and how it operates can be devised from this multiplicity, itself a necessarily incomplete account of the thousands of ways in which K'iche' people in Momostenango interact on a daily basis with what we can call 'the cosmological'.

Read in the perspective of traditional ethnology, this book has introduced some new information that ethnographies of the past were missing, mainly that the 'corral' cosmology found by ethnologists of the sixties and seventies in Chiapas also exists in the Highlands of Guatemala, and that this cosmology can 'explain' the relationship between society and cosmological beings (ancestors and earth-gods) in Momostenango and perhaps elsewhere in Indigenous Guatemala, a topic that lacked a synthesis in past ethnographies. However, at the same time, this book has insisted that the image of Indigenous societies that functionalist ethnology depicted is no longer possible. The most important conclusion of the decades of debates that ensued the collapse of the functionalist model in Mesoamerica and elsewhere is that the aim of anthropology should not be to propose a model of Indigenous societies, but rather understand how Indigenous societies model society and the world. Cosmology is the key term here, but, as mentioned in my introduction, the notion became itself a problematic topic, not so much because of Indigenous ideas, but rather because of our understanding of them.

Indeed, the traditional ethnological representation of 'non-Western' cosmologies as assemblages of symbols that represent (or rather 'distort') reality in an 'ideological' way, a way of representation that dominated Mesoamerican anthropology until recently, is no longer possible, but what then? What I have tried to propose is that, rather than concentrating on complexes of 'symbols', or even general 'ontologies', we should pay attention to the interrelation of elements in Indigenous cosmological thought, and the paradoxes that arise out of their interaction. Specifically, this book has tried to introduce within the anthropological study of cosmologies the idea of 'strange loop', first proposed by the cognitive and computer scientist Douglas Hofstadter (1979, 2007). This idea resides in the relationship of two domains, one containing the other, such as they conform a 'tangled hierarchy', where what is originally conceived as a superior domain with a transitive relationship to the subordinate levels contained within it is suddenly reversed in its role by an ingenious logical device, consisting of a self-referential mapping of one domain into the other: 'A loop's strangeness comes purely from the way in which a system can seem to "engulf itself" through an unexpected twisting around, rudely violating what we have taken to be an inviolable hierarchical order' (Hofstadter 2007: 159).

Indeed, starting from the cosmology of the sacred altars of the K'iche' and working through topics such as personhood, time, causality, activism and history, I hope to have shown that K'iche' cosmology can be conceived as containing similar 'loops' in many domains, thanks to the paradox posited by the sacred shrines and how they contain the universe that contains them, as well as the people who inhabit it, while these people somewhat remain 'outside' of them and can manipulate and interact with them. This allows the K'iche' to modify the cosmological domain that contains them and determine their lives through rituals, affecting the many intersections that cosmology has with the social: family, economy, community and extra-community politics, and more. However, while Hofstadter's idea works through the notion of a symbolic mapping of domains and remains somewhat representational, in this book cosmologies are considered as being actual territories that are imbricated in a 'strangely looping' relationship, and cosmological action and knowledge is considered to become possible by the entanglement of such territories. Thus, the strange loop in K'iche' cosmology is representational to us, but ontological for them.

While it may seem surprising, it is possible to assert that this idea was already anticipated in the most important representatives of Maya ethnology. Earlier in this book, I have mentioned Evon Vogt's notions of 'replication' and 'encapsulation' (1965; 1969: 571–87). The first, as defined by him, means that 'the world of the Zinacantecos is … segmented conceptually in systematic ways that are replicated in different domains of Zinacanteco culture' ([1990] 1970: 130). Indeed, despite his strong functionalist orientation, Vogt already understood how there was a sort of recursive mechanism within the different levels of the political structure, social life and cosmology of the Tzotzil of Zinacantan, a notion that could even be said to anticipate Marilyn Strathern's conceptions on scales and fractality in social anthropology ([1991] 2004). Thus, 'encapsulation' meant for Vogt 'the conceptual and structural incorporation of new elements into existing patterns of social and ritual behaviour'. In an ethnohistorical dimension, for example, it meant that, in Maya cultures, exogenous elements tend to be integrated by being 'surrounded' in native elements, 'encapsulated' by them instead of fused with them or being outright supressed.[1] Central to this idea is his interpretation of the Tzotzil notion of 'embracing': 'There is a stem in Tzotzil, *-pet*, which means "to embrace". The concept of "embracing" occurs in a number of domains in Zinatenteco life: the socialization process in the family, the baptismal ceremonies, the wedding ceremonies, the curing ceremonies, and beliefs about the Ancestral gods inside the mountains' ([1990] 1970: 130). The (functional rather than lexical) K'iche' equivalent of the Tzotzil *pet* is the more textile word *b'otz'ik*, 'to envelop, to wrap'. Indeed, as many ritual practitioners confirmed to me, the main purpose of K'iche' rituals is to *keb'otzik* or 'envelop' people's and even objects' spirits within the sacred space of shrines and altars, thus the many ceremonies in this book can be seen as instances of this idea.

In a similar vein, but within a fully post-functionalist perspective, the ethnologist Pedro Pitarch, who worked among the Tzeltal of Cancuc, Chiapas, has proposed the Deleuzian idea of the fold as a key concept to understand Mesoamerican cosmology in general (2013).[2] He links this idea to the acts of folding and unfolding that take place in numerous Tzotzil rituals, and proposes this as the key to understanding a Mesoamerican 'topology' (2021). Despite going in an opposite direction to Vogt in regards to the latter's functionalist assumptions, it is somewhat ironical that, in the end, Pitarch's ideas can be seen as a sort of development of Vogt's conceptions, but more abstract and complex, emphasizing processes of self-differentiation over those of self-similarity, and more in line with post-functionalist and postmodern preoccupations; that is, concerned with conceptual complexity above social instantiations.

My proposal can be seen as a further step within the same path, attempting to 'close' the circuit (or, rather, create another loop in it) by describing the work of two other related mechanisms that seem to operate in K'iche' cosmology: looping and entanglement. Indeed, the main argument of this book is that K'iche' cosmos reaches into itself, not only iterating itself through mechanisms of self-similarity or fractality across its levels (Vogt's replication reread in a Strathernian fashion), or 'folding' itself through differentiation (Pitarch's Deleuzian folds), but *containing itself* in a way in which that which is contained and that which contains it become entangled in a paradoxical, looping cosmic hierarchy.[3] Thus, the idea presented in this book would be a synthesis of the two main aspects that emerged in the history of Maya ethnological theory; that is, repetition (Vogtian replication), and difference (Pitarchian folding), to speak in Deleuzian terms (Deleuze 1994).

In Momostenango, the cosmos and its inhabitants have external copies of themselves that are identical to the point of having entangled properties, and these copies are 'folded' within the universe via the sacred shrines; however, people also remain different to such copies. This creates the cosmological paradox, which surmounts the perceived difference between the universe and its own image in Western thought. In general Mesoamerican terms, this loop is achieved through the idea of *nahualli*, or *nawal* in K'iche'; in his rereading of Amerindian soul-conceptions, Roy Wagner suggests that the Nahuatl *nahualli* and its correlate opposite outlines a sort of Klein bottle cosmology (2010: 6–7); a similar relationship would be expressed by the opposition between *nawal* and *wach'ib'al* in K'iche'. Thus, the point of the K'iche' shrines is to create a cosmological loop within the K'iche' cosmos, to turn its original disposition upside down.

The possibilities of this paradox have been explored in the preceding chapters: from paradoxes of inclusion to time paradoxes, to paradoxes of causation, to the paradoxes of political and historical action. However, more than a summary of the preceding chapters, the aim of this final section is to try to take these ideas to different levels of discussion, to try to apply this 'cosmological loop' to further theoretical questions. Thus, this final section will discuss two topics: magic and causation, and how cosmology itself can be tentatively conceived.

Magic as a Form of Downward Causation

In many parts of this book, it is possible to find what anthropology has called 'magic', a concept that is problematic by its very definition and remains so.[4] It has been called 'spurious system of natural law' (Frazer [1922] 1990: 11); has been understood as a 'social belief' characterized by ideas regarding 'contiguity', special 'properties' and the agency of 'supernatural beings' (Mauss [1950] 2001); a kind of 'primitive knowledge' that is different from technique and enhances the 'personal perspective' of the native on the performance of technical productions (Malinowski 1948: 15); or analysed as a semiotic transfer of effects (Tambiah 1968, 1973). Nonetheless, it is undeniable that a strange relationship with causality is usually associated with magic, and that this strangeness appears once and again in this book, either to create protection, to alter the outcome of events or even to attack enemies. But what is it?

I have suggested that for a cosmology in itself to be more than a map, a 'frozen' image in the words of Leach (1982: 213), in order for it to be efficient, a 'strange loop' must be achieved. This looping form is not only to be found in the explicit form of K'iche' cosmology, for example, but it is also to be found elsewhere, even if we are missing some of the 'data' that could outline this idea more clearly; for example, in the loop achieved between the earth (*ga*) and the sky (*mu*) in the Kachin seance narrated by Leach (1982: 214), or in the loop suggested to be extant between the domains called Boyowa (the external universe) and Tuma (an invisible, ancestral universe) in the cosmology of the Trobrianders of Kiriwina, for example, which seems to make the magic that Malinowski spoke about possible (Mosko 2014). In general, 'magic' actions imply a certain entanglement of domains that, at first, would seem separate. They also give rise to actions that are not those of the pure social system, or the unchecked cosmological forces in their 'natural flow'. This point is important. Causality for non-Western people remains *always* cosmological, but there is a way in which cosmological causality can be manipulated, it can invert its course and flow from the subordinate domain to the subordinating or 'superior' domain. How to understand this? What I propose, tentatively, is that this sort of paradoxical causality is what Hofstadter, following Roger Sperry (1965) and Donald Campbell (1974), has identified as 'downward causation', defined as 'the causation of lower-level effects by higher-level entities' (Paolini Pauletti and Orilia 2017: 1). But how so, if its definition suggests the opposite of what I have stated?

The notion of downward causation, which was formulated in the study of emergence in biological systems (Campbell 1974), has been extensively debated in philosophy and science since its introduction (cfr. Bøgh Andersen et al. 2000), and has been applied to subjects such as textual criticism (Bøgh Andersen 2011), and sociology (Elder-Vass 2012; Hodgson 2011), where it has been posited as a plausible mechanism of social action. As mentioned, while conceptually articulated by Roger Sperry, the term 'downward causation' was coined by the social and evolutionary scientist Donald T. Campbell and it was defined by him as the way in which 'the laws of [a] higher-level selective system determine, in part, the distribution of lower-level events and substances', or that 'processes at the lower levels of a hierarchy are restrained by and act in conformity to the laws of the higher levels' (1974: 180).

As Dave Elder-Vass considers, at least two different versions of downward causation can be distinguished. The first one would be unproblematic in appearance:

> Perhaps the least problematic version of ontological dependence is that entities of a higher level are composed of entities of a lower level: the parts of a whole are at a lower level than the whole. We may argue, for example, that human individuals are at a higher ontological level than molecules, as they are composed of molecules. Hence, an experiment in chemistry would be a case of this kind of top-down causation, if the human scientists create the conditions in which a particular molecular reaction occurs. (Elder-Vass 2012: 85)

However, there is another version of this idea that becomes more paradoxical, because it implies an interaction of levels within an entity:

> A second, and at first sight more demanding, interpretation of the *top-down* in *top-down causation* is that *top-down* causation occurs only when an entity has a causal impact on its *own* parts. We need to alter our example only slightly to show that much the same considerations apply in this case as in the first interpretation. If our scientist was to conduct her experiment on herself, perhaps by placing a drop of some chemical on her skin and observing/measuring the chemical reaction that ensued, we would have just such a case. The scientist's power to co-cause the chemical reaction remains an emergent power of the scientist as a whole, while the molecules that take part in the reaction, including those that form part of the scientist, also co-cause the reaction. Causation is exerted both by the scientist and by the molecules, and the scientist has a downward causal effect on some of the parts of herself. Events, in other words, may be co-determined by the causal powers of both higher and lower level entities, even when the lower level entities concerned are parts of the higher level entities concerned, and even where the events that result are changes in those same lower level entities and thus also in the higher level entity itself. (Elder-Vass 2012: 85)

Thus, this second version of downward causation implies a 'strange loop'.

As Menno Hulswit remarks, the paradox of downward causation consists in the fact that it contrasts with 'one of the cornerstones of the modern concept of causation ... that a cause must "precede" its effect, so that an effect (for instance, a system) can never *cause its cause* (the constitutive parts of the system)' (2005: 262).[5] In the case of the scientist, the paradox arises when we realize that the scientist is but a mere consequence of the interactions of the neural networks and molecules that constitute him or her, but now exerts a certain sort of causality in relationship to them. In a tentative way, it can be said that magic is a sort of 'reverse-directional cause', to borrow Campbell's biological term (1974: 181), in which an agent exerts causal powers within the very cosmology that defines them and gives them grounding. Thus, the paradox is that, while cosmological entities can be seen as superior entities or domains, in reality they are the lower-level entities, the basis of reality from which causality flows, while people within cosmologies are higher-level entities, meaning that they derive from them. Given the structure of the non-Western cosmologies treated in this book, the possibilities of downward causality become much greater.

As mentioned, downward causality emerged as a non-reductionist model that tried to explain interactions in hierarchical systems of many kinds; for example, those present in biology, where a sort of 'reverse-directional cause' seems to operate in some instances. Since its inception, downward causation has been closely related to the so-called 'emergentist principle' in evolutionary biology; that is, the idea that certain evolutionary laws 'are not described by the laws of physics and inorganic chemistry, and … will not be described by the future substitutes for the present approximations of physics and inorganic chemistry' (Campbell 1974: 180). In a more general way, 'something emerges from some other thing, roughly, if and only if the former is not predictable from the latter' (Paolini Paoletti and Orilia 2017: 2); that is, when it is not possible to reduce an entity to the laws that govern the phenomenon from which it 'emerges'. In consequence, downward causation has sometimes been generally considered as the way in which emergent properties exert causal influence or acquire causal powers (cfr. Heil 2017: 43). Thus, while for us Westerners, cosmologies could be considered as 'emergent' entities that exert this sort of causal power on other societies, what actually emerges in cosmology is not a series of symbols, but the 'I' that works and lives within it. Thus, it could be said that, contrary to us, in the perspective of non-Western people, that which is the background of emergence is cosmology, and that which emerges is themselves. And this leads us to the final concept to be discussed in this book.

A Cosmos Is a Mind-Like Interface

Returning to Hofstadter, one of his obsessions has been to link the 'strange loop' that he found in different representations to the very idea of the 'I', which he developed extensively in his book *I Am a Strange Loop* (2007). He argued that the 'I' is in itself a 'strange loop' in which the natural order of causality is inverted, by which actions are perceived not as arising from the flow of cause and effect of which the mind is a mere consequence, but from the 'I' or mind itself. That is, by all accounts our macroscopic universe is a deterministic flow of causes and effects, even ourselves, our bodies and even actions are like that, but we do not perceive it that way, at least not in what we ascribe to our agency. As Hofstadter asserts: 'there is … a curious upside-downness to our normal human way of perceiving the world: we are built to perceive "big stuff" rather than "small stuff", even though the domain of the tiny seems to be where the actual motors driving reality reside' (Hofstadter 2007: 173). In a reductionistic way, feelings, desires and everything related to the mind, like volition and action, could be said to be caused by the interaction of neurons; that is, low-level interactions. But it is not so for the 'I' or mind. Instead, the mind increasingly perceives itself as the origin of (at least some) feelings, affections and volition, despite the fact that, technically, mind itself is merely the *consequence* of the interaction of its constituent neurons. Thus, the mind fancies itself as the cause of many of the things that happen around it, and in fact becomes this cause in many cases, *even if in the end the mind is just something created by lower-level systems*. An inversion of the flow of causality occurs at some point, and the mind, a simple consequence of highly organized matter and thus theoretically determined by it, becomes the self-perceived source of the flow of cause and effect around it, despite remaining at the bottom of the actual ladder of causality in reductionistic accounts. One could say that cosmology is such a loop, such as actions are perceived as not arising from the usual flow of cause and effect, be it natural or social, but from the interface or loop between world and mind.

Finally, let us reconsider Gregory Bateson's ideas regarding an 'interface' of sorts between 'mind and nature', also called *creatura* and *pleroma* (Bateson and Bateson 1987),[6] which are closely related to what he called the 'ecology of the mind' (1972). In cybernetics, which informed Bateson's views, an interface was originally merely the 'connections between two pieces of electrical equipment', a concept later extended to the connection between such equipment and users, users among users themselves, and more (Krippendorf 1986: 41). In those terms, the contrast between the more traditional idea of world-view and my idea of cosmology would be this:

Worldview: A particular representation of the world, a content of the mind that attempts to 'describe' or map reality.

Cosmology: A mind-like *interface* between the territory of nature and the territory of beings, or *pleroma* and *creatura* in terms of Bateson. A strange loop that makes downward causation possible within self-iterating totalities.

This book suggests that cosmologies are not 'products of the mind' or representations, be them individual or more 'collective', but, rather, culturally oriented 'mindlike' structures that people operate with (both inside and outside them) and somewhat, transiently, assume, different 'I's' in Hofstadter terms. However, the key to understanding such entities is not to conceptualize them as merely emerging products of mental activity, but rather as *the emergent form that mental activity takes by solving the dilemma of totality vis-à-vis its own reflection*. They are 'mind-like' in the sense of the 'I' of Hofstadter; that is, 'strangely looping' structures with 'downward causation' powers, hence, culturally 'emergent' entities. Cosmologies are 'minds' in this sense, 'high-level' entities with an array of powers defined by a tendency to invert the flow of causality. This sense of an action coming from above rather as than from 'below' is what defines true cosmological action. In a way, the cosmological problem lies not within the array of causes and effects than arise from the 'baseline' configuration of the universe, but the way in which such flow is subverted in 'cultural' action. This is what Hofstadter calls 'symbols'. We do not perceive the universe around us as low-level processes, but as organized, high-level 'symbols'. This inversion is the key to the cosmological, *in any culture*, be it Western or Indigenous. At some point, we rebel against the usual flow of what is, and submit it to another flow. This is cosmology: the mind-like interface where natural forces flow according to our own understanding. There is no 'ideology' in this: it is somewhat of the destiny of all high-level entities as soon as they become capable of conceiving a totality that can contain itself in a 'Gödelian' fashion.[7]

Cosmologies are not merely a cultural product of the human mind but, rather, an arrangement of particular 'minds' themselves. They are not identified with the mind of anyone, though, and neither people nor groups are defined by them, rather, people turn to them, take them seriously or play with them, entering and exiting of them depending of the events of their own lives. Cosmologies are not ultimately defined by their contents, but by how they solve concrete problems by solving the problem of totality in itself. Many such cosmologies can coexist in the same person, as we have seen with the dilemmas of 'conversion' in Chapter 1 of this book. Sometimes, cosmologies just sort of lie outside the individual, the individual being indifferent to their working, until they decide to turn to them in order to better their own lives. They are not the 'cultural essence' of a group, but rather something that floats in a social space, and people use them (or not). Cosmologies are thus more like interfaces in Bateson's conception, while also conserving some of the characteristics of the notion of 'I' in Hofstadter. Each of them solves the dilemmas of the universe in a particular way, but to do it they turn the notion of an 'image' of the world 'inside out', and create paradoxical arrangements of 'territories' and their 'maps' (which become territories within them).

Finally, a Myth

The ideas presented formerly could be difficult to understand, but to try to do it in a more figurative fashion, let us analyse by way of conclusion a contemporary K'iche' myth, the myth about the creation of the sacred shrines, which exemplifies the main dilemma of cosmology. Before the altars existed, the world was chaotic, and people had no religion nor rituals. The earth gods became angry because of this, and they started to make fire rain from the sky, in the form of a flammable resin called *q'ol*, which burst into flames when in contact with people's homes. Fearsome animals, the *nawales* of gods, devoured and killed people. Snakes made their nests in people's homes, the jaguars lived in the fields and the monkeys perched atop the roofs, mocking men. Many people escaped, hid underground or in the mountains. Those that did so became animals. Then, in order to stop this, the rest of the survivors 'thought on what to do to be able to defend themselves, to speak to the gods. Then they invented the rituals, they made the shrines for burning and offered the same material (that) was falling from the sky, they offered the same (pine resin) so the gods would not send more, and thus they stopped it and fire and burning resin never fell from the sky again'.

Thus, the important thing in the K'iche' myth is not the existence of the universe or its representation, but rather overcoming the forces within it and their initial dominance over humans. What humanity did was to take the fire that rained out of the sky and turned it into the fire that would burn within their altars. Thus, the cosmological fire in the sky became the ritual fire in their shrines. Through this inversion, causality itself was inverted, and the hellish, apocalyptic former creation became our ordered world. This myth does not only echo with the well-known early colonial mythology of the *Popol Vuh* regarding the destruction of former creations (Christenson 2003: 85–90) and the Aztec myth of the rain of fire that ended one of the earlier cosmological suns (Bierhorst 1992: 142–4), but in a more general way it echoes with

Lévi-Strauss' theory about the 'One myth' presented in the final volume of his *Mythologiques*, where he hypothesizes that the main subject of Amerindian mythology is the transit from celestial fire to domestic fire (1981: 561–624). Thus, the K'iche' myth echoes Lévi-Strauss' theories fully, his contention being that the main topic of Amerindian mythology is the invention of ritual fire from the celestial fire of the gods. But, of course, this idea can be made to go beyond the notion of myth as a mere narrative. In a way, the idea behind Lévi-Strauss' general theory on myth is that 'mythic thought operates essentially through a process of transformation' (1981: 675), a process mainly symbolized by fire in the grand architecture of his magnum opus. However, this process also has a formal correlate: the so-called 'general formula' where 'terms' transform into 'functions'.[8] This can also be seen as another form of the many inversions of levels and hierarchies that this book is about. However, the fire that burns in human culture itself is not merely physical fire, but also the 'fire of the mind'. Cosmology is the conceptual form of that energy which transforms the world, such as myth is its language-form, and fire itself is its physical correlate in myth (and elsewhere). In the end, this is what the K'iche' myth alludes to: that, despite being the same thing, there is a difference between how the universe burns matter and how the mind (or 'culture') burns or transform things within it. This difference soon gives rise to a different cosmos, and the cycle begins anew, in other horizons and other 'creations', as Mesoamericans myths expressed.

Notes

1. Vogt offers many examples of 'encapsulation' of elements external to Zinacantan that, at his time, were new to the town: the Mexican government's schools, new government officials, the Pan-American Highway, and even the Harvard fieldwork house (1970: 582–6). All these elements were integrated into the social life and the cosmology of Zinacantan by performing different rituals in which they were surrounded and somewhat burdened with symbols of native rituals, like candles, flowers, pine needles floors and, in the case of the highway, the erection of multiple shrines with sacred crosses.
2. The notion of fold used by Pitarch is borrowed from Deleuze's famous study of the relationship between Leibniz and baroque art, *The Fold: Leibniz and the Baroque* (1988).
3. It is important to insist here on the difference between fractality (self-similarity across levels) and the theory of the 'strange loop'; that is, structures that are not only self-similar in their organization like fractals but are also self-inclusive, implying with this the possibility of a 'return' to the highest level of their organization from the lower levels, in a paradoxical way whose most important representative are Gödel's incompleteness theorems (cfr. Hofstadter 1979: 5).

4. See Otto and Stausberg 2013 for a history of the definition of the concept in both historical sources and modern social theory.
5. Recently, Menno Hulswit himself (2005), Carl Craver and William Bechtel (2007), as well as Geoffrey M. Hodgson (2011), have called into question the usage of the term 'causation' in regards to the idea of 'downward causation', since, as Craver and Bechtel remark, 'the idea of causation would have to stretch to the breaking point to accommodate interlevel causes' (2007: 573). Hodgson proposes the term 'reconstitutive downward effects' as a substitution. Their criticism and alternatives have not been accepted as a replacement for the concept yet; for example, Elder-Vass in general considers that 'there is no reason why we should follow Craver, Bechtel and Hodgson in thinking of this as merely an effect that is inferior in some way to any other kind of causation' (Elder Vass 2012: 86). I retain through this book the term 'causation' because the kind of downward effects this book deals with are characterized as causal for my informants.
6. Borrowing from a conceptual pair found in Carl Gustav Jung's book *Seven Sermons to the Dead* ([1916] 1982), a pastiche of gnostic treatises, Bateson suggests the existence of two worlds, one called 'Pleroma', and one called 'Creatura'. As explained by Bateson, 'In the Creatura, all is names, maps and names of relations'; however, as he clarifies, 'still the name of the name is not the name, and the name of the relation is not the relation, even when the relation between A and B is of the kind we denote by saying that A is the name of B' (Bateson and Bateson 1987: 21). Conversely, pleroma would be, for Bateson, the physical domain, where things are 'governed by forces and impacts' (Bateson and Bateson 1987: 13). In the Gnostic conception, the Greek word πλήρωμα, which literally means 'fullness' (Liddell and Scott [1901] 1996: 1420), denotes 'the totality of unmanifest qualities of the Godhead, and also at times to those which are manifest in the universe or in man' (Macdermot 1981: 76).
7. While being a very strong inspiration for his ideas, Hofstadter stresses that the importance of Gödel's theorems in his work is not completely related to their content (the incompleteness of any system powerful enough to represent itself, such as mathematics), which have been abused by relativist thinkers in past years (cfr. Franzén 2007), but rather to their form, which achieves, through the usage of a clever numbering, the encoding of a statement about a system as a statement within the system itself: 'Gödel had realized that one of the domains that mathematics can model is the doing of mathematics itself' (Hofstadter 2007: 161). In a way, how cosmology is described to work in this book is similar to this inversion, although not 'derived' from it.
8. For an explanation of Lévi-Strauss' transformational analysis of myth and its relationship with the so-called 'general formula' see Santucci, Doja and Capocchi 2020. For its relationship with cosmological diagrams and 'hourglass configurations', see Lévi-Strauss 2001.

APPENDIX A

Some K'iche' Cosmological Concepts

Calendrics

Since many of the cosmological notions of the K'iche' are crucially defined by their relationship with calendrics, an exposition of K'iche' cosmology must necessarily begin with that topic. Two Indigenous calendar cycles survive among the K'iche': the *cholq'iij*, a sacred cycle of 260 days, and the *junab' q'iij*, the solar cycle of 365 days. Regarding the first, the sacred calendar of the K'iche', called *cholq'iij* ('count of days'), is a calendric cycle formed by the combination of thirteen numerical coefficients and twenty day-signs. A cycle of twenty days with their coefficients would look like this: 1 Imox, 2 Iq', 3 Aq'ab'al, 4 K'at, 5 Kan, 6 Kame, 7 Kej, 8 Q'anil, 9 Toj, 10 T'zi', 11 B'atz', 12 E, 13 Aj, 1 I'x, 2 Tz'ikin, 3 Ajmaq, 4 No'j, 5 Tijax, 6 Kawoq, and 7 Ajpu; the count continues until all the possible combinations of coefficients and day-signs are exhausted, totalling 260 days. This count of days, of deep Mesoamerican roots and of continuous usage in the K'iche' area until today,[1] has no exact correlation to any natural phenomena,[2] being of a ritual nature instead, and is mainly used for ritual and divinatory purposes.

The most important work on the contemporary K'iche' divinatory calendar and its usage in Momostenango is that of Barbara Tedlock, *Time and the Highland Maya* (1992). In a ground-breaking fashion, Tedlock first established that the contemporary divinatory meanings of the day-signs of the K'iche' calendar are not primarily based on the semantics of their names, since many of them are archaic and have lost their meaning in contemporary K'iche', but rather based on a special set of mnemotechnics that relies on paronomasia and wordplay. However, many of these names *still* mean something, and those meanings are often, but not always, related to the divinatory meaning of the day-sign, so it is possible to infer that the mnemotechnics are often a way of preserving ancestral meanings, or even a hermeneutic device of pre-Hispanic origins. In any case, the reader is encouraged to consult Tedlock's pages on the topic to deepen her/his knowledge (1992: 107–31), but a summary of the day-sign meanings is presented here, with some additions of my own. Whenever the meaning of the day-sign's name still has sense, its literal meaning is included alongside the K'iche' word. In general, the concrete value of any day in divinatory readings is heavily determined by its numerical coefficient, the idea being that low numbers mean a low intensity of the *nawal* or spiritual force of the day itself, while high numbers have a stronger connotation. The traditional Mesoamerican order of the calendar has been used here, in order to surmount the many modern disagreements on which day comes first.[3]

1. **Imox:** Denotes insanity and foreigners, from the etymologically unrelated mnemotechnic word *mox*, 'demented' and, thus, metaphorically, 'foreigner'.
2. **Iq':** Associated with wind, often a furious one, thus, it often denotes fury. Curiously, the contemporary word for wind, *kaq'iq'*, is a lexicalization of *kaq'*, 'red, furious' and *iq'*, which was the original root for 'wind'. It is one of the four Year-Bearers, and it denotes a year of bad weather and furious winds.
3. **Aq'ab'al ('Darkness'):** Associated with the idea of dawn and new beginnings.
4. **K'at ('Net'):** It denotes a heavy debt that is carried by the client of a divination as if a net or bundle. It also denotes various transgressions via the mnemotechnic *k'atik*, 'to burn', which is a metaphor for sexual intercourse.
5. **Kan:** The word for this day-sign was 'snake' in many Mesoamerican divinatory calendars and the K'iche' still remember this meaning, although 'snake' is *q'umatz* in their language. It denotes antagonism, vengeance, and is a preferred day for offensive sorcery.
6. **Kame ('Death'):** Associated with death in general. It is good for asking for protection from death, ruin or destruction, and even causing it.

7. **Kiej ('Deer, horse'):** Always associated with power. It is one of the four Year-Bearers, and it denotes a year of political harmony (low coefficients) or political unrest (high coefficients).
8. **Q'anil:** Associated with abundance and fertility, as such, the *winel* shrine is used to make agricultural offerings, and thus is called *awas q'anil*.
9. **Toj ('Payment'):** A good day-sign to makes offerings to apologize or give a 'payment' to the ancestors and earth-gods.
10. **Tz'i ('Dog'):** Associated with dirtiness, impurity, sexual transgression and filth. It also denotes justice, perhaps owing to two associations.
11. **B'atz' ('Thread'):** Associated with weaving, the idea of enveloping (*keb'otz'ik*), and with divinatory knowledge, since 8 B'atz' is the day of the initiation of calendric diviners. For this reason, it is considered in some works as 'the first day' of the divinatory calendar. In colonial times, it was associated with weavers, and probably with Jun B'atz' and Jun Chuwen, the monkey gods of arts.
12. **E':** Associated to the life path in general due to the similarity with the word *b'e*, 'road'. It is one of the four Year-Bearers, and it denotes a year of undertakings and beginnings.
13. **Aj ('Reed'):** Associated with the patrilineage and the family, and, thus, to the *warab'al ja* shrine.[4]
14. **I'x:** The day-sign of the earth. Its Mesoamerican equivalents are related to jaguars, a symbol of the earth among groups like the Tzotzil.
15. **Tz'ikin ('Bird'):** Associated with money and with the sacred altar called *meb'il*.
16. **Ajmaq ('Sinner'):** Associated with asking for forgiveness because of transgressions. It is also the day-sign of dead people in general, although not of divinized ancestors.
17. **No'j ('Understanding'):** Associated with wisdom. It is one of the four Year-Bearers, and it denotes a year of general understanding among the community.
18. **Tijax ('Flint'):** Associated with medicine, with war and strife (*choj*), with authoritative words (*katb'al tz'ij* or 'knife words') and even with slander (*xak'ab'al chi'aj*).
19. **Kawok:** Associated with altars and tables filled with offerings via the mnemotechnic *remelik tikilik*, 'sitting and gathered'. A good day-sign to make an offering.
20. **Ajpu ('Blowgunner'):** The day of the divinised dead ancestors, the *rawajal ri uwachulew*.

In regards to the solar year, or *junab' q'iij*, the cycle has 365 days with no leap days, meaning that the beginning of the count of the year shifts one day back every four years in relationship to the Gregorian Calendar. For example, during my stay in town, I witnessed the celebration of the new K'iche' year on the eve of 19 February 2018, the year starting on 20th; however, in 2025, the K'iche' year starts on 19 February. By virtue of the mechanics of the Mesoamerican calendar, each year can only begin on four of the day-signs of the *cholq'iij*: Iq', Kiej, E' and No'j, each day having one of thirteen numerical coefficients. This day-sign is called a 'year bearer' since, as it will be further explained in the next section, each year is associated with one of the four cosmological deities known as the Mam. In a similar way to people, each year (or its Mam, more precisely) has a character defined by its Nawal or calendric sign. For example, 2017 was a 5 Iq' ('Wind') year, and, as such, the strong rains and winds present in the hurricane season were attributed to it; in 2018, a 6 No'j ('Wisdom') year, it was expected that peace and understanding would prevail in the community. In general, Iq' is associated with winds and storms, Kiej to political turmoil, E' to a good year to start any project, and No'j to understanding and peace.

Altars and Ritual Geography

Momosteco ritual geography is defined by four sacred shrines that rest on mountains located in each of the cardinal points. According to local mythology, in the origin of the world, four aged but strong divinities known as Mam ('grandfathers', aged spirits that bear the mountains on their backs with a tumpline, although they are sometimes seen as just one spirit) carried these mountains on their backs and placed them on each of the four directions at particular dates of the Maya calendar, starting with the mountain of the east at the day Kiej. Each one of these mountains (Kilaja, Tamanku, Sokob' and Pipil Ab'aj, going clockwise, associated with Kiej, Iq', E' and No'j, respectively) is the seat of one of the four Mam or year-bearer gods that rule each year. The day-sign of the current Mam is considered to be the *alcalde* or ruler among the day lord gods and each one of them has a subordinate day lord that acts as his 'secretary' or second in command; these 'secretaries' have special shrines of their own. Each year, the main ritual calendric specialist of the town performs a ceremony to 'welcome' the ruling Mam at its corresponding mountain shrine. Generally speaking, ceremonies that affect the whole town are associated with these mountain shrines too.

Altars in the immediate vicinity of the town also have a complex ritual geography, which has been treated in other works (Van der Akker 2018: 29–32, 43–48; Tedlock 1992: 101). These altars are the seats of the different members of the 'bureaucracy' of the Holy Earth. Gobierno Mundo ('Government of the Earth') seems to be located in San Francisco el Alto, far from Momostenango, and is associated with days numbered 12 and 13. Presidente Mundo ('President of the Earth') is located at Nima' Mesab'al ('Big Broom'), and it is also known as B'elejeb'al 'the place of number 9'. Ministro Mundo ('Minister of the Earth') is located at the altar known as Chuti Mesab'al ('Little Broom'), also known as Wajxaqib'al, 'the place of number 8'. The next-highest-ranking altar is the Pak'lom Hill, which is considered a cosmological nexus and the centre of the world; it is known as Waq'ib'al, 'the place of number 6'. Secretario Mundo seems to be Pa Ja' ('By the Waters') a river altar known as Junib'al ('The Place of Number 1'). Finally, other two altars are of great importance. One is associated with the patron saint of the town, Santiago, and it is known as Puja'l Santiago ('The Spring of Santiago'); the other is Kokoch ('Archway'), an altar in front of the main Catholic church of the town, associated with the dead. These altars integrate a ritual circuit that all diviners must visit upon initiation.[5]

Soul and Body Conceptions

The following is a summary of an article of mine on the topic (Zamora Corona 2019). There are four main 'soul' conceptions or 'co-essences' (Monaghan 1998) among the K'iche' of Momostenango, which can also be found in the glossary.

1. ***K'ux:*** A word meaning 'heart', the *k'ux* is a 'soul' with human form. It is said to be deposited in the ancestral altars of the K'iche' families, especially those called *winel* and *warab'al ja*. It is also called *anima*.
2. ***Chikop:*** A soul with the form of a little bird, representing the most vulnerable part of a person, which can be attacked by angry earth (*mundo*) spirits, or sorcerers.
3. ***Nawal:*** Unlike the *k'ux*, the *nawal* is a soul with animal form. It is said to correspond to an actual animal living in the mountains; thus, when such an animal is killed, the person dies. It also defines the personality of a person: a person with an eagle *nawal* is daring, a person with a coyote *nawal* is cunning, and so

on. *Nawal* also refers to two more things: the first is the agency of a person or animal conceived as a 'heat-force'; thus, a deadlier animal has a strong, bigger and hotter *nawal* than a harmless one; for example, a venomous snake has a greater *nawal* than a cow despite their respective size. Finally, *nawal* also refers to the calendric birth-sign of a person, which is also called *wach uq'iij* or 'the visage of someone's day'.

4. ***Win:*** An animal or spectral spirit invoked by a sorcerer by performing 'black magic' ceremonies with the specific aim to damage or even kill another. The word is also used to designate such a sorcerer. This co-essence does not belong to all people, only to sorcerers, although, according to some testimonies I found, in the past, men of prominent political power also had it and used it to protect the town from hostile *win* spirits. The *win* can be obtained using a number of complex rituals, and, in fact, some people are said to have magic books where the prayers and procedures of such rituals are detailed.

Besides these soul conceptions, Indigenous ideas on the human body in Momostenango also have a number of features that are different from those found in Western conceptions. The human body in its entirety is called, in a metonymical fashion, *baq'il* ('bone') or sometimes *tiojil* ('flesh'). The corporeal presence of a person is called *wachib'al*, but the term is also applied to images or depictions of people, saints and spirits. Finally, the blood of a person (*kik'el*), can have an extra 'soul' called *koyopa* ('lightning'), which can receive messages during calendrical divination.

Divination

While K'iche' rituality is multifaceted and many rituals are explained in this book, a word is needed here in regards to calendrical divination, to which Barbara Tedlock dedicated her most important work (1992), which is not described in detail in the preceding chapters. K'iche' calendrical divination (*ch'obonik* or 'understanding') is performed by calendrical diviners (male and female) through the counting of red beans of the coral tree (*Erythrina corallodendron*), called *tz'ite'* in K'iche'. A calendrical diviner has a bundle (*b'otz'ik* or *vara*) where they keep 260 beans, each corresponding to a day in the K'iche' calendar, as well as crystals called *chok'*, which represent the hierarchy of the year bearers, their secretaries, and which are variable in number (ten in the canonical arrangement reported by Tedlock, sometimes just one, and sometimes even more crystals mixed with the red beans). In divination, a calendric priest will invoke the cosmological forces, untie his divining bag, spread the mixture of beans and crystals on the table, then line up the ten crystals representing the highest cosmological hierarchy, and take a handful of the beans and crystals, putting aside the reminder. The seeds of this handful will be arranged in fours, until there are no more (the final arrangement must be two or one, never three). Each one of these beans or crystals corresponds to one day of the Maya calendar (Figure 8.1).

The diviner counts them, starting from the day where the divination is carried out, sometimes backwards into the past, sometimes forwards into the future, depending on the question. At many points during the count, the diviner will feel a sting, twitch or pulse in her/his blood or body, the *koyopa*, the 'lightning soul' mentioned above. This is called *kacha' ukik'el* ('his/her blood speaks'), and this indicates to the diviner which day wants to speak with him/her, as well as its message, according partly to a set of corporeal symbolisms (see Tedlock 1992: 141 for some of them), and partly to direct communication of the day with the diviner. For example, in a very simple divination consultation carried out by my friend Rudy on my behalf, I had dreamed of a friend the night prior and wanted to know the meaning of the dream. The day that spoke during the dream was Tzik'in, the *nawal* of money. What I dreamed was not my friend, but the *nawal* of money, which took the 'visage' (*wach*)

Figure 8.1. Rudy Ramos, performing *ch'ob'onik* with his divining bundle. © Alonso Zamora Corona

of my friend in the dream, and the diviner concluded that his message was that I would have some transient economic difficulties, which to my surprise (and not surprisingly for him) occurred not long after the divining session, since I had some problems with my card in Guatemalan banks, which caused a temporary setback to my finances during the week that they took to resolve. Thus, divination can answer simple life questions as the one explained in the preceding lines, it can establish the proper time for rituals or determine such complex decisions such as the succession of the main calendric priest of the town, as I have explained elsewhere in this book.

Notes

1. See Weeks, Sachse and Prager (2013) for a general account and historical antecedents.
2. See Rice 2007: 31–39 for an overview of the problem. Three explanations on the signification of the number 260 have been stated: that the cycle corresponds to the average span of human pregnancy (280 days), some aspects of the synodic cycle of Venus (about 263 days of apparition as the Morning Star and a similar period as the Evening Star), or the cycle of maize agriculture in the Highlands (for example, in 2018, I helped with planting maize on 24 February and the harvest was on 23 November, approximately 272 days of interval). The K'iche' strongly favour the first hypothesis, despite not being completely exact.
3. Tedlock considered that the count has no beginning (1992: 96); in general, many contemporary diviners consider the day B'atz' as the first, 8 B'atz' being the day of the initiation of all diviners. However, since the calendar derives from Mesoamerican traditions, which are unanimous in starting the cycle with the equivalent days Imix (Yucatec), Cipactli (Nahuatl), Quevui (Mixtec), Chilla (Zapotec), the first day-sign here will be its K'iche' counterpart, Imox.
4. There is a certain confusion in certain sources regarding the meaning of this day, given the similarity in sound and orthography between the words *aj*, 'cane', and *aj*, 'corncob'. This confusion happens because the current Maya orthography does not differentiate this pair of vowels, which are actually different (cfr. Pontious 1980: v). This pair of words would be written *aj* and *äj* in the old *indigenista* alphabet.
5. See Van der Akker 2018: 46 for a map of the aforementioned places.

APPENDIX B

Glossary

***Ajaw ri uwachulew* ('Lord of the face of the world'):** A term denoting gods and divinised ancestors. The spirits that rule over the forces of nature are also called *rajawal* ('lords') or *dueños* ('owners', in Spanish).

***Ajitz*:** Witch or sorcerer, he who performs *itzinel* or witchcraft. See also *win*.

***Ajq'iij* ('He of the day'):** A calendrical diviner, not necessarily the head of a patrilineage.

***Alaxik* ('Birth'):** A K'iche' patrilineage.

***Auxiliatura* ('Auxiliary government'):** A Spanish word, denoting the seat of the local traditional government, elected by its former members or *pasados*. There are local *auxiliaturas* and a town *auxiliatura* or Indigenous *alcaldía*. It is usually formed by a major, a secretary, a treasurer and their replacements, as well as different assistants.

***Awas* ('Taboo'):** A word denoting sacred objects, things that cannot be named directly and ritual transgressions. Usually, it denotes the sacred shrines made of piles of stones and potsherds where offerings are burned to honour earth gods and ancestors. Altars dedicated to ancestors form a hierarchy of sorts, similar to that of ritual specialists or *chuchqajaw*: there are *awas rech alaxik* or patrilineage shrines, *awas rech kanton* or community shrines, and the *awas rech tinamit*, the town shrine. Also known as *porob'al* or *quemadero* in Spanish (both words meaning 'burning places').

***Awajmundo* ('Earth-gods animal'):** The animal spirits under the command of the earth gods, and the animal guise of the earth gods themselves, both of which are fearsome and can deal punishment to humans who commit transgressions against the earth gods.

***Chikop* ('Animal, bird'):** One of the souls or 'co-essences' (Monaghan 1998) of a person. It takes the form of a little bird. It is completely defenceless, and can be attacked by angry gods, ancestors and sorcerers. It is analogous to the 'bird of the heart' of the Tzotzil (Guiteras 1961: 270) and Tzeltal (Pitarch 2010: 22–24).

***Chuchqajaw* ('Priest-shaman'):** The head of a *costumbrista* patrilineage, responsible for both familiar and ritual functions, but also a name for Indigenous ritual specialists in charge of an altar or *awas*. There is a ritual hierarchy of sorts among them: there are *chuchqajaw* for a patrilineage, for a locality and for the whole town. Being highly honorific, the preferred term nowadays is *kamal b'e* ('guide of the road').

***Dioses Mundo*:** Earth gods. Also called *e rajawal* 'owners'. As a single entity they are called Dios Mundo, in singular.

***Juan No'j*:** One of the patron gods of sorcerers. He lives at the Santa María Volcano in Xelajuj Noj (Quezaltenango). Originally a sorcerer too, after his death he became the lord or *rajawal* of the volcano. He exploits souls that toil in a supernatural hacienda that is constantly burning. He is also the master of aggressive *awaj-mundo* who serve him and other sorcerers.

***Kationik* ('Bark and bite'):** Both the warning and the action of punishment that ancestors and gods perform on those who commit ritual transgressions.

***K'oxol*:** A dwarf spirit, guardian of treasures, owner of game and the patron of divination. Often dressed in red and thus called *k'aki' k'oxol*.

***Koyopa* ('Sheet lightning'):** A tingle or feeling in the body of a diviner that gives a message to him/her, depending on where it was felt.

***K'ux* ('Heart'):** One of the souls or 'co-essences' of a person. It has a human form and it is the closest to the Western idea of a soul or spirit. Also called *anima*.

***Mam* ('Grandfather'):** Earth spirits visualized as old men carrying the mountains with a tumpline. There are four chiefs among them, associated with four mountains to each of the cardinal points, as well as the calendrical signs that rule the solar year.

***Maximon*:** An Indigenous saint figure representing Judas, whose cult emerged in Santiago Atitlán. His name is a compound of *Mam Ximon*, 'grandfather Simón'. His figure is revered and then discarded and burned during the Holy Week. He is possibly the modern counterpart of the ancient god Rilaj Mam (Vallejo Reina 2005), and perhaps of the Classic Maya god Chuwaaj (Beliaev and Tokovinine 2013: 184–91), an aged god of commerce, prosperity and wealth, very fond of smoking, drinking and of women.

***May*:** A calendrical period of twenty years. It also means the former generations or cohorts of the dead. They are said to be thirteen in total. In some prayers they are associated with thirteen major patrilineages and thirteen ideal divisions of the land. A synonymous expression is *le'* ('line, rank').

***Meb'il*:** One of the three kind of shrines that constitute the *awas*. It can be a stone with the form of an animal, or even a box with precious stones or crystals inside. Dedicated to financial success and wealth increase. Its associated day is Tz'ikin.

***Mesa* ('Table'):** A grouping of altars dedicated to the earth gods. Often shaped like dolmens or stone tables, sometimes featuring crosses. They do not belong to particular patrilineages, but have a ritual specialist in charge of them, the *aj mesa*. Its associated calendrical day is K'at.

Muni**:** The municipal government building, the former barracks of Teodoro Cifuentes, military strongman during the dictatorships of Estrada Cabrera and Jorge Ubico.

Nantat **('Mothers and fathers'):** The deceased and often divinized ancestors of a patrilineage.

Nawal **('Hidden, supernatural being'):** One of the souls or 'co-essences' of a person. It represents two things: the calendrical sign of a person, and a companion animal spirit that is said to live and die in parallel to the person. Also called *wach uq'iij* ('the visage of his day'). The term also refers to the power or agency of a person conceived as 'heat-force' or *q'aq'il*.

Nawal Mesa**:** A ritual specialist, often called 'spiritualist' in the extant literature (Tedlock 1992: 74). He has a direct communication with the earth gods and ancestors and can be possessed by them, performing séances. There are three kinds among them: green, red and black. The black is the strongest and deals with 'forgotten ancestors', as well as the hungriest and most aggressive among the earth gods.

San Antonio Mundo**:** A *costumbrista* saint, the patron of rural patrilineages. He is called 'the saint of the holy earth' and has the form of a child. Partly inspired by the catholic San Antonio de Padua, who in some images carries baby Jesus. San Antonio Mundo is the protector of the animal spirits or *nawales* of families and therefore is also called 'pastor of the holy earth'.

Santiago**:** The *costumbrista* patron saint of the town. He is called 'captain of the holy earth' and is represented as a Guatemalan soldier of the nineteenth century on a horse, wielding a sabre. Partly inspired by Santiago Apóstol, patron saint of the conquistadors. Associated with the morning star, he controls the legions of the elements and is deeply related with political power.

Santo Mundo **('Holy Earth'):** This expression either designates the earth divinities conceived as a single being, or the cosmological world known as *Juyub' Taq'aj* ('mountain valley). It also comprises the sky or *Kaj*, thus, in it, above and below are equivalent.

Tew kaqiq' **('Cold and wind'):** An expression somewhat synonymous with the 'spiritual'. Ancestors and gods are said to be present *pa tew kaqiq'*, 'in the cold and the wind' when rituals are performed in their honour. It is also similar to the term *uxlab'* or 'breath'.

Wachib'al **(Lit. 'Face-hood'):** The image and bodily presence of a person, extending to any icon or symbol of it. It can have agency, as is the case of the images of saints.

Wajxaqib B'atz' **('8 Thread'):** The most important calendric celebration in Momostenango. It consists of a ritual circuit of offerings beginning in the altar at the Chuti Mesab'al hill, then Pa Ja', then Uja'l Santiago, then Pak'lom Hill, then a prayer at the main church at Momostenango and finally at the Kokoch altar before the daybreak. The circuit culminates at the next day, at Nima Mesab'al.

Warab'al ja **('House of dreaming'):** One of the three kinds of altars that constitute the *awas*. Dedicated to the cult of the ancestors. Its associated day is Aj.

Win: A hostile spirit used by a sorcerer to do harm against his victims. It can be animal or spectral, and they are usually carnivorous. The sorcerer can either send this spirit to people, or transform into it through different rituals. By extension, a sorcerer who can command this kind of spirits is called *win* too. A *win* sorcerer must be born on the day Tijax. Days associated with the *win* are Iq, Kan, Kame, Kiej, Toj, Kiej, Ix, depending on the form of the spirit called.

Winel: One of the three kind of altars that constitute the *awas*. Dedicated to the cult of fertility. Its associated day is Q'anil.

References

Aburto Delgado, Leonel. 2019. 'La biografía del señor presidente: sujetos políticos y racialización en *¡Ecce Pericles!* de Rafael Arévalo Martínez', *Revista Ístmica* 23: 29–42.

Abramson, Allen, and Martin Holbraad. 2014. 'Introduction: The Cosmological Frame in Anthropology', in Allen Abramson and Martin Holbraad (eds), *Framing Cosmologies: The Anthropology of Worlds*. Manchester: Manchester University Press, pp. 1–30.

Adams, Richard. 1952. *Un análisis de las creencias y prácticas médicas en un pueblo indígena de Guatemala: Con sugerencias relacionadas con la práctica de medicina en el área maya*. Guatemala: Instituto Indigenista Nacional.

Adams, Richard. 1968. *Crucifixion by Power: Essays on Guatemalan National Social Structure, 1944–1966*. Austin TX: University of Texas Press.

Adas, Michael, Peter Stearns and Stuart B. Schwarz. 2009. *Turbulent Passage: A Global History of the Twentieth Century*. Upper Saddle River NJ: Pearson Education.

Aguirre Beltrán, Gonzalo. 1967. *Regiones de refugio: El desarrollo de la comunidad y el proceso dominical en Mestizoamérica*. Mexico: Instituto Nacional Indigenista.

Akker, Paul van der. 2018. 'Time, History and Ritual in a K'iche' Community: Contemporary Maya Calendar Knowledge and Practices in the Highlands of Guatemala', PhD diss., Leiden University.

Albores Zárate, Andrea. 1978. *El funcionalismo en la etnografía tzeltal-tzotzil: un análisis de sus implicaciones teóricas y políticas*. Tuxtla Gutiérrez, Mexico: Universidad Autónoma de Chiapas.

Anderson-Imbert, '"Magic Realism" in Spanish American Fiction', *The International Fiction Review* 1(1): 1–7.

Arévalo Martínez, Rafael. 1945. *¡Ecce Pericles!* Guatemala: Tipografía Nacional.

Arifín-Cabo, Pressia. 2011. *Prácticas Tradicionales Maya de Resolución de Conflictos en los Territorios K'iche', Tzutujil y Kaqchikel.* Guatemala: Deutsche Gesellschaft für Internationale Zusammenarbeit. https://www.ziviler-friedensdienst.org/sites/default/files/media/file/2022/zfd-practicas-tradicionales-maya-de-resolucion-de-conflictos-1531_8.pdf (Accessed 20 September 2024).

Asturias, Miguel Ángel. [1946] 2000. *El señor presidente.* Ed. by Gerald Martin. Paris: ALLCA XX.

Ball, Phillip. 2018. *Beyond Weird: Why Everything You Thought You Knew about Quantum Physics Is Different.* Chicago IL: University of Chicago Press.

Banaji, Jairus. 1970. 'The Crisis of British Anthropology', *New Left Review*, 0(64): 71–85.

Barnard, Alan. 2000. *History and Theory in Anthropology.* Cambridge: Cambridge University Press.

Bateson, Gregory. 1958. *Naven: A Survey of the Problems Suggested by a Composite Picture of the Culture of a New Guinea Tribe Drawn from Three Points of View.* Stanford CA: Stanford University Press.

Bateson, Gregory. 1972. *Steps Towards an Ecology of the Mind.* Chicago IL: The University of Chicago Press.

Bateson, Gregory 1979. *Mind and Nature: A Necessary Unity.* New York: E. P. Dutton.

Bateson, Gregory and Mary Bateson. 1987. *Angels Fear: Towards an Epistemology of the Sacred.* New York: Macmillan Publishing.

Batz, Leopoldo. 2015. 'Momostenango: defender el bien común es suficiente para ser criminalizado', *Prensa Comunitaria*, 2 July. https://comunitariapress.wordpress.com/2015/07/02/momostenango-defender-el-bien-comun-es-suficiente-para-ser-criminalizado/ (Accessed 30 September 2024).

Bayly, Susan. 2018. 'Anthropology and History', in Mattei Candea (ed.), *Schools and Styles of Anthropological Theory.* London: Routledge, pp. 108–20.

BBC News. 2019. 'El gobierno de Guatemala ordena la expulsión de la CICIG'. *BBC News*, 8 January. https://www.bbc.com/mundo/noticias-america-latina-46790805 (Accessed 30 September 2024).

Beliaev, Dmitri, and Alexandre Tokovinine. 2013. 'People of the Road: Traders and Travelers in Ancient Maya Words and Images', in Kenneth G. Hirth and Joanne Pillsbury (eds), *Merchants, Markets, and Exchange in the Pre-Columbian World.* Washington D C, Dumbarton Oaks, pp. 169–200.

Bell, Elizabeth R. 2012. 'Sacred Inheritance: Cultural Resistance and Contemporary Kaqchikel-Maya Spiritual Practices', PhD diss., Ohio State University.

Bertrán, Pedro. 1897. 'Los indios que se redimen: Cartas de estudiantes indígenas aventajados', *La Ilustración Guatemalteca* I(11): 168–70.

Bierhorst, John. 1992. *History and Mythology of the Aztecs: The Codex Chimalpopoca.* Tucson AZ: The University of Arizona Press.

Bocek, Barbara. 2006. 'Everyday Politics in a K'iche' Village of Totonicapán, Guatemala', in Walter Little and Timothy Smith (eds), *Mayas in Postwar Guatemala: Harvest of Violence Revisited.* Tuscaloosa AL: The University of Alabama Press, pp. 124–38.

Bøgh Andersen, Peter, et al. (eds). 2000. *Downward Causation. Minds, Bodies and Matter.* Århus: Aarhus University Press.

Braudel, Fernand. 1958. 'Histoire et Sciences sociales: La longue durée', *Annales* 13(4): 725–53.

Bright, William. 1990. '"With One Lip, with Two Lips": Parallelism in Nahuatl', *Language* 66(3): 437–52.

Brintnall, Douglas E. 1979. *Revolt Against the Dead: The Modernization of a Mayan Community in the Highlands of Guatemala*. New York: Gordon and Breach.

Cámara Nueva Vision. 2014. *Elección de Autoridades Ancestrales de Momostenango*. DVD. Momostenango: Cámara Nueva Visión.

Campbell, Donald T. 1974. '"Downward Causation" in Hierarchically Organised Biological Systems', in Francisco Jose Ayala and Theodosius Grigorievich Dobzhansky (eds), *Studies in the Philosophy of Biology: Reduction and Related Problems*. Berkeley CA: University of California Press, pp. 179–86.

Cancian, Frank. 1976. *Economía y prestigio en una comunidad maya*. México: Instituto Nacional Indigenista.

Cancian, Frank. 1994. *The Decline of Community in Zinacantan: Economy, Public Life, and Social Stratification, 1960–1987*. Stanford CA: Stanford University Press.

Capponi, Anna Sulai. 2006. 'El culto de Santiago entre las comunidades indígenas de Hispanoamérica: símbolo de comprensión, reinterpretación y compenetración de una nueva realidad espiritual', *Imaginario* 12(13): 249–77.

Cardoza y Aragón, Luis, and Manuel Arce. 1984. 'La tiranía inefable o el sustento mágico del poder: "Brujos y soldados de Momostenango lo guardaban"', *Kanina: Revista de Artes y Letras de la Universidad de Costa Rica* 8(1–2): 33–38.

Carlsen, Robert S. 2011. *The War for the Heart and Soul of a Highland Maya Town*. Revised ed. Austin TX: University of Texas Press.

Carmack, Robert. 1972. 'Ethnohistory: A Review of Its Development, Definitions, Methods, and Aims', *Annual Review of Anthropology* 1: 227–46.

Carmack, Robert. 1979. *Historia social de los quichés*. Guatemala: Editorial 'José de Pineda Ibarra', Ministerio de Educación.

Carmack, Robert. 1981. *The Quiché Mayas of Utatlan: The Evolution of a Highland Guatemala Kingdom*. Norman OK: The University of Oklahoma Press.

Carmack, Robert (ed.). 1988. *Harvest of Violence: The Maya Indians and the Guatemalan Crisis*. Norman OK: University of Oklahoma Press.

Carmack, Robert. 1995. *Rebels of Highland Guatemala: The Quiché-Mayas of Momostenango*. Norman OK: The University of Oklahoma Press.

Carmack, Robert. 1998. 'Traditional Momostenango: A Microhistoric Perspective on Maya Settlements Patterns, Political Systems and Ritual', in Andrés Ciudad Ruiz (ed.), *Anatomía de una civilización: aproximaciones interdisciplinarias a la cultura maya*. Madrid: Sociedad Española de Estudios Mayas, pp. 323–52.

Caso, Alfonso. 1958. *Indigenismo*. Mexico: Instituto Nacional Indigenista.

Centro Latinoamericano y Caribeño de Demografía. 2019. 'XII Censo Nacional de Población y VII de Vivienda. Principales Resultados Censo 2018', Guatemala: Instituto Nacional de Estadística. https://www.ces.uvg.edu.gt/censo2018/documentos/Principales_resultados_Censo2018.pdf (Accessed 30 September 2024).

Cerezo Arévalo, Marco Vinicio. 'Acuerdo Gubernativo No. 1046–87 de noviembre 23 de 1987'. http://mayaglot.com/blog/wp-content/uploads/2016/05/ACUERDO-GUBERNATIVO-ALFABETO-MAYA.pdf (Accessed 30 September 2024).

CERIGUA (Centro de Reportes Informativos sobre Guatemala). 2011. 'Lider indígena fue atacado por encapuchados'. *La Hora*, 4 November 2011, 8.

Chandler, David. 2011. *A Dictionary of Media and Communication*. Oxford: Oxford University Press.

Chiquin, Cristina. 2013. 'Momostenango se dispone defender su territorio'. *Prensa comunitaria*, 29 May. http://comunitariapress.blogspot.com/2013/05/momostenango-se-dispone-defender-su.html (Accessed 30 September 2024).

Christenson, Allen J. 2003. *Popol Vuh: The Sacred Book of the Maya*. Winchester and New York: O Books.

Comisión para el Esclarecimiento Histórico. 1999. *Guatemala: Memoria del Silencio*. Guatemala: Oficina de Servicios para Proyectos de las Naciones Unidas.

Consejo Departamental de Autoridades Ancestrales y Comunitarias. 2016. 'Propuesta de Reforma Constitucional sobre Reconocimiento de la Jurisdicción Indígena presentada por el Consejo Departamental de Autoridades Ancestrales y Comunitarias del Departamento de Totonicapán', *Hacia la Reforma de la Justicia en Guatemala*. https://www.reformajusticia.org.gt/wp-content/uploads/2016/06/002.pdf (Accessed 17 September 2020).

Cook, Garrett W. 1981. 'Supernaturalism, Cosmos and Cosmology in Quichean Expressive Culture', PhD diss., State University of New York.

Cook, Garrett W. 2000. *Renewing the Maya World: Expressive Culture in a Highland Town*. Austin TX: University of Texas Press.

Cook, Garrett, and Thomas Offit. 2008. 'Ritual Symbols in the K'iche' Tutelary Deity Complex'. Foundation for the Advancement of Mesoamerican Studies, Research Report. http://www.famsi.org/reports/07075/index.html (Accessed 13 September 2024).

Cook, Garrett, and Thomas Offit. 2013. *Indigenous Religion and Cultural Performance in the New Maya World*. Albuquerque NM: University of New Mexico Press.

Couffon, Claude. 1962. 'Miguel Angel Asturias et le réalisme magique', *Les Lettres Françaises* 954: 6.

Craver, Carl F., and William Bechtel. 2007. 'Top-Down Causation Without Top-Down Causes', *Biology and Philosophy* 22(4): 547–63.

de la Cadena, Marisol. 2015. *Earth Beings: Ecologies of Practice Across Andean Worlds*. Durham NC: Duke University Press.

de Smit, Bart, and Lenstra, Hendrik. 2003. 'The Mathematical Structure of Escher's Print Gallery', *Notices of the AMS* 50(4): 446–51.

de Smit, Bart. 2003. 'Animations of the Filled in Print', in *Escher and the Droste Effect*. https://www.math.leidenuniv.nl/~desmit/escherdroste/page_menu=animation.html (Accessed 30 September 2024).

Deleuze, Gilles. 1993. *The Fold: Leibniz and the Baroque*. London: The Athlone Press.

Deleuze, Gilles. 1994. *Difference and Repetition*. London: The Athlone Press.

Descola, Philippe. 1986. *La nature domestique. Symbolisme et praxis dans l'écologie des Achuar*. Paris: Éditions de la Maison des sciences de l'homme.

Descola, Philippe. 2013. *Beyond Nature and Culture*, trans. Janet Lloyd. Chicago IL: University of Chicago Press.

Díaz, Ana. 2020. 'Introduction: Rethinking the Mesoamerican Cosmos', in Ana Díaz (ed.) *Reshaping the World: Debates on Mesoamerican Cosmologies*. Boulder CO: University of Colorado Press, pp. 3–30.

Dobroski, Sonja. 2019. 'Time Behaves Badly: Fractal Mathematics for an Anthropology of Time', Paper presented at ASA19: Anthropological Perspectives on Global Challenges, 1 February 2019.

Domínguez, Edgar. 2013. 'Comunidades de Momostenango realizan consulta por minería', *Prensa Libre*. 1 September. https://www.prensalibre.com/ciudades/totonicapan/comunidades-momostenango-realizan-consulta-mineria-0-985101681/ (Accessed 30 September 2024).

Domínguez, Edgar. 2015a. 'Pobladores exigen renuncia de alcalde de Momostenango', *PrensaLibre*. 16 June. https://www.prensalibre.com/ciudades/totonicapan/pobladores-exigen-renuncia-de-alcalde-de-momostenango/ (Accessed 30 September 2024).

Domínguez, Edgar. 2015b. 'Autoridades ancestrales y ediles alcanzan acuerdos', *Prensa Libre*. 1 July. https://www.prensalibre.com/ciudades/totonicapan/autoridades-ancestrales-y-ediles-alcanzan-acuerdos/ (Accessed 30 September 2024).

Domínguez, Edgar 2016. 'Comienza juicio contra exalcalde sindicado de discriminación', *Prensa Libre*. 16 August. https://www.prensalibre.com/ciudades/totonicapan/comienza-juicio-contra-exalcalde-sindicado-de-discriminacion/ (Accessed 30 September 2024).

Donovan, Patricia. 2005. 'Shattering Shaman Myths: Tedlock's New Book Explores Female Roots of Shamanism', *UB Reporter*, 7 April. http://www.buffalo.edu/ubreporter/archive/vol36/vol36n28/articles/Tedlock.html (Accessed 30 September 2024).

Dubuisson, Daniel. 2019. *The Invention of Religions*, trans. Martha Cunningham. Sheffield: Equinox.

Durán, Fray Diego. 1971. *Book of the Gods and Rites and the Ancient Calendar*, trans. Fernando Horcasitas and Doris Heyden (eds). Norman OK: University of Oklahoma Press.

Elder-Vass, Dave. 2012. 'Top-Down Causation and Social Structures', *Interface Focus* 2(1): 82–90.

Estrada Peña, Canek. 2010. '*Ri juyub', ri q'aq', ri nuch* (El cerro, el fuego, el bebé): Acerca de la presentación de un niño k'iche' ante el Mundo', *Estudios Mesoamericanos. Revista del programa de Posgrado en Estudios Mesoamericanos* 9: 35–62.

Estrada Peña, Canek. 2013. 'Ideas del tiempo cíclico en la cuenta de 260 días entre los k'iche' contemporáneos. El ciclo ritual del día 8 B'atz', in Mercedes de la Garza and María del Carmen Valverde (eds), *Continuidad, cambios y rupturas en la religión maya*. Mexico: Universidad Nacional Autónoma de México, Instituto de Investigaciones Filológicas, Centro de Estudios Mayas, pp. 215–55.

Estrada Peña, Canek. 2015. '*Anima' ri cho, anima' ri plo* (espíritu de la laguna, espíritu del mar): acerca de la significación del día Imox entre los k'iche', *Estudios de Cultura Maya* 45: 191–219.

Evans-Pritchard, Edward E. 1940. *The Nuer: A Description of the Modes of Livelihood and Political Institutions of a Nilotic People.* Oxford: Clarendon Press.
Evans-Pritchard, Edward E. 1939. 'Nuer Time-Reckoning', *Africa: Journal of the International African Institute* 12(2): 189–216.
Evans-Pritchard, Edward E. 1961. *Anthropology and History.* Manchester: Manchester University Press.
Evans-Pritchard, Eduard E. [1937] 1976. *Witchcraft, Oracles and Magic Among the Azande.* Oxford: Oxford University Press.
EZLN (Ejército Zapatista de Liberación Nacional). 1994. 'Al pueblo de México: hablaron los hombres verdaderos, los sin rostro. Mandar obedeciendo. Subcomandante Marcos', in *Memoria Política de México*, 26 February. http://www.memoriapoliticademexico.org/Textos/7CRumbo/1994-Mandar_obedeciendo.html (Accessed 30 September 2024).
Favret-Saada, Jeanne. 1980. *Deadly Words: Witchcraft in the Bocage.* Cambridge: Cambridge University Press.
Feuchtgang, Stephan. 1992. *The Imperial Metaphor: Popular Religion in China.* New York: Routledge.
Feuchtgang, Stephan. 2014. 'Coordinates of Body and Place: Chinese Practices of Centring', in Allen Abramson and Martin Holbraad (eds), *Framing Cosmologies: The Anthropology of Worlds.* Manchester: Manchester University Press, pp. 116–33.
Figuerola Pujol, Helios. 2010. *Los hombres, los dioses y las palabras en la comunidad de San Juan Evangelista Cancuc en Chiapas.* Mexico: Universidad Nacional Autónoma de México.
Fitzgerald, Timothy. 2000. *The Ideology of Religious Studies.* New York: Oxford University Press.
Flores Escalante, Aída, and Enrique Kuny Mena. 2004. *Tomás Regalado, el último caudillo de Cuscatlán.* San Salvador: El Salvador.
FODIGUA. 1991. *Comunidad étnica K'iche'. Historia y memorias. Volumen II.* Guatemala: Universidad Rafael Landívar.
Fogelson, Raymond D. 1974. 'On the Varieties of Indian History: Sequoyah and Traveller Bird', *Journal of Ethnic Studies* 2: 105–12.
Fogelson, Raymond D. 1984. 'Who Were the Ani-kutani? An Excursion into Cherokee Historical Thought', *Ethnohistory* 31(4): 255–63.
Fogelson, Raymond D. 1989. 'The Ethnohistory of Events and Nonevents', *Ethnohistory* 36(2): 133–47.
Franzén, Torkel. 2007. *Gödel's Theorem: An Incomplete Guide to Its Use and Abuse.* Wellesley, MA: A. K. Peters/CRC Press.
Frazer, James George. [1922] 1990. *The Golden Bough: A Study in Magic and Religion.* London: Palgrave MacMillan.
Garay, Alejandro. 2014. 'El problema de la escritura prehispánica K'iche' vista a través de las referencias y evidencias documentales coloniales', *Estudios Digital* 2: 2–40.
García Escobar, Carlos René. 2007. *Los convites en Guatemala.* Guatemala: Centro de Estudios Folklóricos, Universidad de San Carlos de Guatemala.

García Mora, Carlos, and Andrés Medina, (eds). 1986. *La quiebra política de la antropología social en México.* Mexico: Universidad Nacional Autónoma de México.
Geertz, Clifford. 1966. *Person, Time and Conduct in Bali: An Essay in Cultural Analysis.* New Haven: Yale University Press.
Gell, Alfred. 1992. *The Anthropology of Time: Cultural Constructions of Temporal Maps and Images.* Oxford: Berg Publishers.
Gershman, Boris. 2022. 'Witchcraft Beliefs Around the World: An Exploratory Analysis', *PLoS One* 17(11): e0276872.
Gobaud Carrera, Antonio. 1937. *The Guajxaquip Báts: An Indian Ceremony of Guatemala.* Guatemala: Centro Editorial 'José de Pineda Ibarra', Ministerio de Educación Pública.
Gobaud Carrera, Antonio. 1964. *Indigenismo en Guatemala.* Guatemala: Centro Editorial 'José de Pineda Ibarra', Ministerio de Educación Pública.
Godínez Orantes, Lester Homero. 2015. *La Marimba. Un estudio histórico, organológico y cultural.* Mexico: Fondo de Cultura Económica.
Gossen, Gary H. 1975. 'Animal Souls and Human Destiny in Chamula', *Man* 10(3): 448–61.
Grecco Pacheco, Daniel. 2021. 'The Concept of Wrapping and Its Ontological Character among the Maya', *Estudios Latinoamericanos* 41: 47–65.
Griaule, Marcel. 1948. *Dieu d'eau. Entretiens avec Ogotemmêli.* Paris: Les Editions du Chêne.
Guatemala Friendship School Foundation. 2010. 'About the GFSF', *The Guatemala Friendship School Foundation*, 29 July. http://gfsfschool.blogspot.com/p/about-gfsf.html (Accessed 30 September 2024).
Guernsey, Julia, and F. Kent Reilly. 2006. *Sacred Bundles: Ritual Acts of Wrapping and Binding in Mesoamerica.* Barnardsville NC: Boundary End Archaeology Research Center.
Guiteras Holmes, Calixta. 1961. *Perils of the Soul: The World View of a Tzotzil Indian.* New York: Free Press of Glencoe.
Hage, Ghassan. 2012. 'Critical Anthropological Thought and the Radical Political Imaginary Today', *Critique of Anthropology* 32(3): 285–308.
Hallowell, Alfred Iriving. 1964. 'Ojibwa ontology, behavior, and world view', in S. Diamond (ed.), *Cultural History: Essays in Honor of Paul Radin.* New York: Octagon Books, pp. 19–51.
Hansson Wahlberg, Tobias. 2013. 'Dissolving McTaggart's Paradox', in Christer Svennerlind, Jan Almäng and Rögnvaldur Ingthorsson (eds), *Johanssonian Investigations.* Berlin, Boston: De Gruyter, pp. 240–58.
Hart, Thomas. 2008. *The Ancient Spirituality of the Modern Maya.* Albuquerque NM: University of New Mexico Press.
Hegelund, Allan. 2005. 'Objectivity and Subjectivity in the Ethnographic Method', *Qualitative Health Research*, 15(5): 647–68.
Heil, John. 2017. *Downward Causation.* London: Routledge.
Hermitte, M. Esther. 1970. *Poder sobrenatural y control social.* México: Instituto Indigenista Interamericano.

Historia de Momostenango (blog). 2018. 'Reforma de Momostenango' n/d. http://juyubuleuw.blogspot.com/p/reforma-de-momostenango.html (Accessed 30 September 2024).

Hobsbawm, Eric, and Terence Ranger (eds). 1983. *The Invention of Tradition.* Cambridge: Cambridge University Press.

Hodgson, Geoffrey M. 2011. 'Downward Causation: Some Second Thoughts', in *Geoffrey M Hodgson.* https://www.geoffreymhodgson.uk/downward-causation (Accessed 30 September 2024).

Hoenes del Pinal, Eric. 2022. *Guarded by Two Jaguars: A Catholic Parish Divided by Language and Faith.* Tucson AZ: University of Arizona Press.

Hofstadter, Douglas. 1979. *Gödel, Escher, Bach: An Eternal Golden Braid.* New York: Basic Books.

Hofstadter, Douglas. 2007. *I Am a Strange Loop.* New York: Basic Books.

Holbraad, Martin. 2012. *Truth in Motion: The Recursive Anthropology of Cuban Divination.* Chicago: University of Chicago Press.

Holbraad, Martin. 2020. 'The Shapes of Relations: Anthropology as Conceptual Morphology', *Philosophy of the Social Sciences* 50(6): 495–522.

Holbraad, Martin, and Morten Axel Pedersen. 2017. *The Ontological Turn: An Anthropological Exposition.* Cambridge: Cambridge University Press.

Holland, William R. 1961. 'El tonalismo y el nagualismo entre los tzotziles', *Estudios de Cultura Maya* 1: 167–81.

Holland, William R. 1964. *Medicina Maya en los Altos de Chiapas.* Mexico: Instituto Nacional Indigenista.

Howell, Signe. 2012. 'Cosmology', in Alan Barnard and Jonathan Spencer (eds), *The Routledge Encyclopedia of Social and Cultural Anthropology.* Second edition. London: Routledge, pp. 157–60.

Hübner, Manuel Eduardo. 1992. *Guatemala en la Historia. Un pueblo que se resiste a morir.* Guatemala: Foto Publicaciones.

Hudson, Charles. 1966. 'Folk History and Ethnohistory', *Ethnohistory* 13: 52–70.

Hugh Farmer, David. 2004. *The Oxford Dictionary of Saints.* Oxford: Oxford University Press.

Hulswit, Menno. 2005. 'How Causal is Downward Causation?' *Journal for the General Philosophy of Science* 36(2): 261–87.

Ingthorsson, Rögnvaldur. 2016. *McTaggart's Paradox.* London: Routledge.

Jung, Carl Gustav. [1916] 1982. *Seven Sermons to the Dead*, trans. Stephan A. Hoeller. Wheaton IL; Chenai, India: Quest Books.

Kapferer, Bruce. 2002. 'Introduction: Outside All Reason – Magic, Sorcery and Epistemology in Anthropology', in *Beyond Rationalism Rethinking Magic, Witchcraft and Sorcery.* New York/Oxford: Berghahn Books, pp. 1–30.

Kaputsa, Jan. 2021. 'Maya Crosses Dressed, Fed, Alive. A Case Study on Maya Animism', *Ibero-Americana Pragensia* 49(1): 53–68.

Katovich, Michael A. 1987. 'Durkheim's Macrofoundations of Time: An Assessment and Critique', *The Sociological Quarterly* 28(3): 367–85.

Killick, Evan. 2015. 'Perspectives on Climate Change and Its Mitigation: Ontological Wars in Amazonia', keynote paper given at the conference, *Trans-Environmental Dynamics: Understanding and Debating Ontologies, Politics, and History in Latin*

America, LMU Munich October 2015. https://www.academia.edu/20759277/Perspectives_on_Climate_Change_and_its_Mitigation_Ontological_Wars_in_Amazonia (Accessed 30 September 2024).

Kirchhoff, Paul. 1943. 'Mesoamérica: Sus límites geográficos, composición étnica y caracteres culturales', *Acta Americana* 1: 92–107.

Klein, George Linam. 1990. 'The 'Prophetic Perfect', *Journal of Northwest Semitic Languages* 16: 45–60.

Knowlton, Timothy. 2002. 'Diphrastic Kennings in Mayan Hieroglyphic Literature', *Mexicon* 24(1): 9–13.

Korsbaek, Leif. 1990 'La religión y la política de cargos: una comparación de tres comunidades mayas en los Altos de Chiapas', *Cuicuilco* 23–24: 115–29.

Korsbaek, Leif. 2017. 'La historia y la antropología: El sistema de cargos', *CIENCIA ergo-sum* 2(2): 175–83.

Korzybski, Alfred. 2000 [1933]. 'A Non-Aristotelian System and its Necessity for Rigour in Mathematics and Physics', in *Science and Sanity: An Introduction to Non-Aristotelian Systems and General Semantics*. New York: Institute of General Semantics, pp. 747–62.

Krech III, Shepard. 1991. 'The State of Ethnohistory', *Annual Review of Anthropology* 20: 345–75.

Krippendorf, Klaus. 1986. *A Dictionary of Cybernetics*. Philadelphia PA: The Annenberg School of Communications, University of Pennsylvania.

La Farge, Oliver. 1947. *Santa Eulalia: The Religion of a Cuchumatan Indian Town*. Chicago IL: University of Chicago Press.

La Farge, Oliver, and Douglas Byers. 1931. *The Year Bearer's People*. Publication no. 3. Middle American Research Institute. New Orleans LA: Tulane University.

Lamrani, Miriam. 2008. 'Nagualism and Tonalism: An Insight on Human/Non-Human Relationships in Mesoamerica', MA diss., University College London.

Lamrani, Miriam. 2018. 'From the Saints to the State: Modes of Intimate Devotion to Saintly Images in Oaxaca, Mexico', PhD diss., University College London.

Latour, Bruno. 2004. *Politics of Nature: How to Bring the Sciences into Democracy*. Cambridge MA; London: Harvard University Press.

Leach, Edmund R. 1950. 'Primitive Calendars', *Oceania* 20: 245–62.

Leach, Edmund. 1971. *Rethinking Anthropology*. London: The Athlone Press, University of London.

Leach, Edmund R. 1982. *Social Anthropology*. Glasgow: Fontana.

Lévi-Strauss, Claude. (1962) 1966. *The Savage Mind*. Chicago IL: University of Chicago Press.

Lévi-Strauss, Claude. 1981. *The Naked Man. Introduction to a Science of Mythology 4*, trans. John and Doreen Weightman. New York: Harper and Row.

Lévi-Strauss, Claude. 2001. 'Hourglass Configurations', in Pierre Maranda (ed.), *The Double Twist: From Ethnography to Morphodynamics*. Toronto: University of Toronto Press, pp. 15–32.

Lewis, Stephen E. 2018. *Rethinking Mexican Indigenismo: The INI's Coordinating Center in Highland Chiapas and the Fate of a Utopian Project*. Albuquerque NM: University of New Mexico Press.

Leys, Jos. 2007. 'The Droste Effect Image Transformation', *Computers & Graphics* 31(3): 516–23.

Liddell, Henry George, and Robert Scott. [1901] 1996. *A Greek English Lexicon*. Ninth edition with a revised supplement, revised and augmented by Sir Henry Stuart Jones and Roderick McKenzie. Oxford: Clarendon Press.

López Austin, Alfredo. 1988. *The Human Body and Ideology: Concepts of the Ancient Nahuas*, trans. Thelma Ortiz de Montellano, and Bernardo Ortiz de Montellano. Salt Lake City UT: University of Utah Press.

López Austin, Alfredo. 1990. *Los mitos del tlacuache*. México: Alianza Editorial Mexicana.

López Austin, Alfredo. 2018. 'Sobre el concepto de cosmovisión', in *Juego de Tiempos*. Mexico: Academia Mexicana de la Lengua, pp. 66–99.

Macdermot, Violet. 1981. 'The Concept of Pleroma in Gnosticism', in *Gnosis and Gnosticism*. Leyden: Brill, pp. 76–81.

Mackenzie, James. 2016. *Indigenous Bodies, Maya Minds: Religion and Modernity in a Transnational K'iche' Community*. Denver CO: University of Colorado Press.

MacMillen, Christian. 2009. *Making Indian Law: The Hualapai Land Case and the Birth of Ethnohistory*. New Haven CT: Yale University Press.

Malinowski, Bronislaw. 1927. 'Lunar and Seasonal Calendars in the Trobriands', *Journal of the Anthropological Institute of Great Britain and Ireland* 57: 203–15.

Malinowski, Bronislaw. 1948. *Magic, Science and Religion and Other Essays*. Boston MA: The Beacon Press.

Malinowski, Bronislaw. [1922] 2002. *Argonauts of the Western Pacific*. London: Routledge.

Marroquín Gálvez, Ernesto. 2014. 'Historia del Cementerio General Santa Isabel: Personajes ilustres, infraestructura y sucesos trascendentales como valor patrimonial y cultural del municipio de Santa Ana', *Anuario de Investigación* 3: 15–36.

Martínez González, Roberto. 2016. *El nahualismo*. Mexico: Universidad Nacional Autónoma de México, Instituto de Investigaciones Antropológicas.

Mauss, Marcel. [1950] 2001. *A General Theory of Magic*. London: Routledge.

McBryde, F. W. 1947. *Cultural and Historical Geography of Southwest Guatemala*. Institute of Social Anthropology, Pub. 4. Washington DC: Smithsonian Institute.

McDaniel, Kris. 2020. 'John M. E. McTaggart', *The Stanford Encyclopedia of Philosophy*. https://plato.stanford.edu/archives/sum2020/entries/mctaggart/ (Accessed 24 September 2024).

McTaggart, J. E. 1908. 'The Unreality of Time', *Mind* 17: 456–73.

McTaggart, J. E. 1927. *The Nature of Existence*. 2 volumes. Cambridge: Cambridge University Press.

Mellor, D. H. 1998. *Real Time II*. Cambridge: Cambridge University Press.

Mendelson, Michael. 1957. *Religion and World View in a Guatemalan Village*. Microfilm Collection of Manuscripts on Middle American Cultural Anthropology, no. 52. Chicago IL: University of Chicago Library.

Mendelson, Michael. 1965. *Los escándalos de Maximón*. Publication no. 19. Guatemala City: Seminario de Integración Social Guatemalteca.

Méndez Bauer, María Belén. 2015. 'Arqueología, historia e identidad en la ciudad de Quetzaltenango, Guatemala', MA thesis, Universidad Nacional Autónoma de México.

Milbrath, Susan. 2000. *Star Gods of the Maya: Astronomy in Art, Folklore and Calendars.* Austin TX: University of Texas Press.

Millán, Saúl. 2015. 'Ontologías en fuga: a propósito de un artículo de Miguel Bartolomé', Trace 67: 158–62.

Millán, Saúl. 2019. 'The Domestication of Souls: A Comparative Approach to Mesoamerican Shamanism', *Social Analysis* 63(1): 64–82.

Monaghan, John. 1998. 'The Person, Destiny, and the Construction of Difference in Mesoamerica', *RES: Anthropology and Aesthetics* 33: 137–46.

Mondragón, Carlos. 2015. 'Recursividad 2.0'. Conference at the Institute of Anthropological Research, Universidad Nacional Autónoma de México.

Morales, Henry (ed.). 2014. *Extractivismo y Derechos Humanos: Legislación, Fiscalidad, Organización y Género.* Guatemala: Cholsamaj.

Morales Sic, José Roberto. 2004. 'Religión y política: El proceso de institucionalización de la espiritualidad en el Movimiento Maya Guatemalteco', MA diss., Guatemala: Facultad Latinoamericana de Ciencias Sociales.

Mosko, Mark. 2014. 'Malinowski Magical Puzzles: Towards a New Theory of Magic and Procreation in Trobriand Society', *HAU: Journal of Ethnographic Theory* 4(1): 1–47.

Muntz, Peter. 1956. 'History and Myth', *The Philosophical Quarterly* 6 (22): 1–16.

National Council of Indigenous Ancestral Authorities of Guatemala. 2009. 'Jutiapa, Guatemala: Queda Establecido Consejo Nacional de Autoridades Ancestrales Indígenas de Guatemala', *Serjus: Asociación Comunitaria para el Desarrollo.* http://www.serjus.org.gt/nota%20autoridades.html (Accessed 18 October 2019).

Neurath, Johannes. 2008. 'Etnografía, epistemología y el proyecto de una antropología teóricamente dialógica', *Diario de Campo* 97: 52–63.

Neurath, Johannes. 2023. *Las religiones indígenas de Mesoamérica: Historia, ritos y transformaciones.* Buenos Aires: SB Editorial.

Niehaus, Isak. 2012. 'Witchcraft and Sorcery', in Alan Barnard and Jonathan Spencer (eds), *The Routledge Encyclopedia of Social and Cultural Anthropology*. Second edition. London: Routledge, pp. 714–17.

Nowotny, Karl Anton. 1976. *Codex Borgia: Biblioteca Apostolica Vaticana (Messicano Riserva 28).* Graz: Akademische Druck-u. Verlagsanst.

Obeyesekere, Gananath. 1992. *The Apotheosis of Captain Cook: European Mythmaking in the Pacific.* Princeton NJ: Princeton University Press.

Observatorio de Conflictos Mineros de América Latina. 2024. 'Conflicto Minero: Comunidades de Momostenango se oponen a proyectos mineros en su territorio'. *Observatorio de Conflictos Mineros de América Latina, Base de datos versión 2.4.5.* https://mapa.conflictosmineros.net/ocmal_db-v2/conflicto/view/960 (Accessed 26 January 2024).

Otto, Bernd-Christian, and Michael Stausberg. 2013. *Defining Magic: A Reader.* London: Routledge.

Paolini Paoletti, Michele, and Francesco Orilia. 2017. 'Downward Causation: An Opinionated Introduction', in Michele Paolini Paoletti and Francesco Orilia (eds), *Philosophical and Scientific Perspectives on Downward Causation*. New York: Routledge, pp. 1–21.

Pedersen, Morten Axel. 2011. *Not Quite Shamans: Spirit Worlds and Political Lives in Northern Mongolia*. Ithaca NY: Cornell University Press.

Pedersen, Morten Axel. 2017. 'Transitional Cosmologies: Shamanism and Postsocialism in Northern Mongolia', in Allen Abramson and Martin Holbraad (eds), *Framing Cosmologies: The Anthropology of Worlds*. Manchester: Manchester University Press, pp. 164–81.

Pérez de Lara, Olga. 1993. 'El desarrollo de la antropología en Guatemala: Necesidades y perspectivas', *Cuadernos de Antropología* 9: 15–35.

Pitarch Pedro. 2000. 'Conjeturas sobre la identidad de los santos tzeltales', *Journal de la Société des Americanistes* 86: 129–48.

Pitarch, Pedro. 2010. *The Jaguar and the Priest: An Ethnography of Tzeltal Souls*. Mexico: Fondo de Cultura Económica.

Pitarch, Pedro. 2013. *La cara oculta del pliegue: antropología indígena*. Mexico: Artes de México.

Pitarch Pedro. 2020. 'Seis cualidades de una etnografía experimental mesoamericana', in *Mesoamérica. Ensayos de etnografía teórica*. Madrid: Nola Editores, pp. 7–29.

Pitarch, Pedro. 2021. 'The Folds of the World: An Essay on Mesoamerican Textile Topology', *HAU: Journal of Ethnographic Theory* 11(3): 1191–205.

Ponciano, Karen. 2007. 'Mayanización y experiencia religiosa: una lectura a partir de los aportes etnográficos sobre la espiritualidad maya', in Santiago Bastos and Aura Cumes (eds), *Mayanización y vida cotidiana. La ideología multicultural en la sociedad guatemalteca*. Volume 3. Guatemala: Cholsamaj, pp. 283–305.

Pontious, David Henne, comp. 1980. *Diccionario Quiché-Español*. Researched by Abraham García Hernández, Santiago Yac Sam. Guatemala: Summer Institute of Linguistics.

Porres, Juan. 1992. 'Los brujos de Samayac', *Estudios: Revista de Antropología, Arqueología e Historia* 1(92): 27–62.

Povinelli, Elizabeth. 2016. *Geontologies: A Requiem to Late Liberalism*. Duke NC: Duke University Press.

Povinelli, Elizabeth. 2021. *Between Gaia and Ground: Four Axioms of Existence and the Ancestral Catastrophe of Late Liberalism*. Duke NC: Duke University Press.

Pozas Arciniegas, Ricardo. 1974. *Los indios en las clases sociales de México*. Mexico: Siglo XXI.

Questa Rebolledo, Alessandro. 2017. 'Dancing Spirits. Towards a Masewal Ecology of Interdependence in the Northern Highlands of Puebla, Mexico', PhD diss., Graduate School of Arts and Sciences, University of Virginia.

Questa Rebolledo, Alessandro. 2023. *Montañas que danzan. Laboratorios maseual para el mantenimiento del mundo*. Buenos Aires: SB Editorial.

Rambo, Terry. 1977. 'Closed Corporate and Open Peasant Communities: Reopening a Hastily Shut Case', *Comparative Studies in Society and History* 19(2): 179–88.

Redfield, Robert. 1941. *The Folk Culture of Yucatan*. Chicago IL: University of Chicago Press.

Redfield, Robert. 1952. 'The Primitive World View', *Proceedings of the American Philosophical Society* 96: 30–36.

Rendón, Mary Catherine. 1988. 'Manuel Estrada Cabrera, Guatemalan President 1898–1920', PhD diss., Oxford: Merton College.

Reynoso, Carlos. 2006. *Complejidad y caos: Una exploración antropológica*. Buenos Aires: SB Editorial.

Rice, Prudence. 2007. *Maya Calendar Origins: Monuments, Mythistory, and the Materialization of Time*. Austin TX: University of Texas Press.

Rice, Prudence. 2013. 'Texts and the Cities: Modeling Maya Political Organization', *Current Anthropology* 54(6): 684–715.

Rio, Knut, and Annelin Eriksen. 2014. 'A New Man: The Cosmological Horizons of Development, Curses, and Personhood in Vanuatu', in Allen Abramson and Martin Holbraad (eds), *Framing Cosmologies: The Anthropology of Worlds*. Manchester: Manchester University Press, pp. 55–76.

Rivera Andía, Juan Javier and Cecilie Vindal Ødegaard. 2019. 'Introduction: Indigenous Peoples, Extractivism, and Turbulences in South America', in Juan Javier Rivera Andía and Cecilie Vindal Ødegaard (eds), *Indigenous Life Projects and Extractivism: Ethnographies from South America*. Cham, Switzerland: Palgrave MacMillan, pp. 1–50.

Robles Castellanos, José Fernando. 2007. *Culhua Mexico: Una revisión arqueo-histórica del imperio mexica tenochca*. Mexico: Instituto Nacional de Antropología e Historia.

Rosaldo, Renato. 1980. *Ilongot Headhunting, 1883–1974: A Study in Society and History*. Stanford CA: Stanford University Press.

Rousseau, Phillipe. 1975. 'Structure and Event in Anthropology and History', *New Zealand Journal of History* 9: 22–40.

Sahlins, Marshall. 1985. *Islands of History*. Chicago IL: The University of Chicago Press.

Sahlins, Marshall. 1999. 'Two or Three Things that I Know about Culture', *The Journal of the Royal Anthropological Institute* 5(3): 399–421.

Saler, Benson. 1960. 'The Road from El Palmar: Change, Continuity, and Conservatism in a Quiché Community', PhD diss., University of Pennsylvania.

Saler, Benson. 1964. 'Nagual, Witch, and Sorcerer in a Quiché Village', Ethnology 3(3): 305–28.

Saler, Benson. 1977. 'Supernatural as a Western Category', *Ethos* 5(1): 31–53.

Saler, Benson. 1993. *Conceptualizing Religion: Immanent Anthropologists, Transcendent Natives, and Unbounded Categories*. Leiden: E. J. Brill.

Santucci, Jean-Francois, Albert Doja, and Laurent Capocchi. 2020. 'Discrete-Event Simulation of Claude Lévi-Strauss' Structural Analysis of Myths Based on Symmetry and Double Twist Transformations', *Symmetry* 12(10): 1706.

SAT (Superintendencia de Administración Tributaria). 2012. 'Impuesto Único Sobre Inmuebles. El ABC de los impuestos'. https://contasociedadesprof.wordpress.com/wp-content/uploads/2017/07/abc-de-los-impuestos.pdf (Accessed 30 September 2024).

Saunders, Clare. 2022. 'Activism', in Clare Saunders et al. (eds), *The Wiley-Blackwell Encyclopedia of Social and Political Movements*. Oxford: John Wiley & Sons, pp. 1–3.

Schultze-Jena, Leonhart. 1933. *Indiana I. Leben, Glaube und Sprache der Quiché von Guatemala*. Jena: Gustav Fischer.

Scott, Charles, and Nancy Tuana. 2017. '*Nepantla*: Writing (from) the In-Between', *The Journal of Speculative Philosophy* 31(1): 1–15.

SERJUS. 2012. 'Procesos Territoriales', *Serjus: Asociación Comunitaria para el Desarrollo*, n/d. http://moremaca.tripod.com/procesos.htm (Accessed 15 September 2019).

SERJUS. 2014. *El sistema de cargos en la autoridad comunitaria y ancestral en el proceso de reconstitución del pueblo maya*. Guatemala: Servicios Jurídicos y Sociales.

Sharer Robert, and Loa Traxler. 2006. *The Ancient Maya*. Palo Alto CA: Stanford University Press.

Shorter, Michael. 1986. 'Subjective and Objective Time', *Proceedings of the Aristotelian Society, Supplementary Volumes* 60: 223–34.

Smith, Quentin. 1994. 'The Logical Structure of the Debate about McTaggart's Paradox', in L. Nathan Oaklander and Quentin Smith (eds), *The New Theory of Time*. New Haven CT: Yale University Press, pp. 202–10.

Sorokina, Anna. 2017. 'Why Did a Young Russian Doctor Open a Hospital in a Remote Guatemalan Village?' Russia Beyond, 4 December. https://www.rbth.com/lifestyle/326918-russian-doctor-opens-hospital-guatemala (Accessed 27 September 2020).

Sperry, Roger W. 1965. 'Mind, Brain and Humanist Values', in J. R. Platt (ed.), *New Views of the Nature of Man*. Chicago IL: University of Chicago Press, pp. 71–92.

Sperry, Roger W. 1976. 'Mental Phenomena as Causal Determinants in Brain Function', in Gordon G. Globus, Grover Maxwell and Irwin Savodnik (eds), *Consciousness and the Brain: A Scientific and Philosophical Inquiry*. New York; London: Plenum, pp. 163–77.

Stengers, Isabelle. 2005. 'The Cosmopolitical Proposal', in Bruno Latour y Peter Weibel (eds), *Making Things Public*. Cambridge MA: MIT Press, pp. 994–1003.

Stengers, Isabelle. 2010. *Cosmopolitics I*, trans. Robert Bononno. Minneapolis MN: University of Minnesota Press.

Stoll, David. 1991. *Is Latin America Turning Protestant? The Politics of Evangelical Growth*. Berkeley CA: University of California Press.

Strathern, Marilyn. [1990] 2013. 'Artifacts of History: Events and the Interpretation of Images', in *Learning to See in Melanesia*. Manchester: HAU Society for Ethnographic Theory, pp. 157–78.

Strathern, Marilyn. [1991] 2004. *Partial Connections*. Updated edition. Oxford: Altamira Press.

Strathern, Marilyn. 1999. *Property, Substance and Effect: Anthropological Essays on Persons and Things*. London: The Athlone Press.

Stuart, David. 1995. 'A Study on Maya Inscriptions', PhD diss., Nashville: Vanderbilt University.

Stuart, David. 2012. 'The Name of Paper: The Mythology of Crowning and Royal Nomenclature on Palenque's Palace Tablet', *Maya Archaeology* 2: 116–41.

Sturtevant, William C. 1966. 'Anthropology, History, and Ethnohistory', *Ethnohistory* 13(1/2): 1–51.

Tambiah, Stanley J. 1968. 'The Magical Power of Words', *Man* 3: 175–208.

Tambiah, Stanley J. 1973. 'Form and Meaning of Magical Acts: A Point of View', in Robin Horton, and Ruth Finnegan (eds), *Modes of Thought. Essays on Thinking in Western and Non-Western Societies*. London: Faber & Faber, pp. 228–9.

Taube, Ronda. 2013. 'Foreign Characters: Visualizing Identity in the Guatemalan Highlands in the Twenty-First Century', in Garrett Cook, and Thomas Offit (eds), *Indigenous Religion and Cultural Performance in the New Maya World*. Albuquerque NM: University of New Mexico Press, pp. 96–118.

Tax, Sol. 1941. 'World View and Social Relations in Guatemala', *American Anthropologist* 43(1): 27–42.

Tax, Sol. 1953. *Penny Capitalism: A Guatemalan Indian Economy*. Chicago IL: University of Chicago Press.

Tedlock, Barbara. 1978. *Quiché Maya Divination: A Theory of Practice*. Albany NY: State University of New York and Albany.

Tedlock, Barbara. 1986. 'On a Mountain Road in the Dark: Encounters with the Quiché Maya Culture Hero', in Gary Gossen (ed.), *Symbol and Meaning Beyond the Closed Community: Essays in Mesoamerican Ideas*. Albany NY: Institute for Mesoamerican Studies, The University at Albany, State University of New York, pp. 125–38.

Tedlock, Barbara. 1992. *Time and the Highland Maya*. Nuevo México: University of New Mexico Press.

Tedlock, Barbara. 2005a. *The Woman in the Shaman's Body: Reclaiming the Feminine in Religion and Medicine*. New York: Bantam Books.

Tedlock, Barbara. 2005b. 'About Barbara Tedlock'. http://www.mayasites.com/btedlock.html (Accessed 3 February 2025).

Turner, Victor. 1975. *Revelation and Divination in Ndembu Ritual*. Ithaca NY: Cornell University Press.

United Nations Development Programme. 2009. *Assessment of Development Results: Guatemala*. New York: United Nations Development Programme Evaluation Office.

Vallejo Reina, Alberto. 2005. *Por los caminos de los antiguos nawales: Rilaj Maam y el nawalismo maya tz'utujil en Santiago Atitlán*. Mexico: Instituto National de Antropología e Historia.

Vela Castañeda, Manolo. 2014. *Los pelotones de la muerte: la construcción de los perpetradores del genocidio Guatemalteco*. Mexico: El Colegio de México.

Véliz Estrada, Rodrigo. 2015. 'El Extractivismo en Guatemala', in Lilián González Guevara (ed.), *El extractivismo en América Central: Un balance del desarrollo de las industrias extractivas y sus principales impactos en los países centroamericanos*. Panamá: Fundación Friedrich Ebert/FES América Central, pp. 59–84.

Villa Rojas, Alfonso. 1963. 'El nagualismo como recurso de control social entre los grupos mayances de Chiapas, México', *Estudios de Cultura Maya* 3: 243–60.

Villa Rojas, Alfonso. 1978. *Los elegidos de Dios. Etnografía de los Mayas de Quintana Roo*. Mexico: Instituto Nacional Indigenista.

Viqueira Albán, Juan Pedro. 2002. *Encrucijadas chiapanecas: economía, religión e identidades*. El Colegio de México.

Viveiros de Castro, Eduardo. 1992. *From the Enemy's Point of View: Humanity and Divinity in an Amazonian Society*. Chicago IL: University of Chicago Press.

Viveiros de Castro, Eduardo. 1998. 'Cosmological Deixis and Amerindian Perspectivism', *The Journal of the Royal Anthropological Institute* 4(3): 469–88.

Viveiros de Castro, Eduardo. 2018. 'The Problem of Affinity in Amazonia', trans. Giovanni da Col, *HAU: Journal of Ethnographic Theory* 8(1–2): 347–93.

Vogt, Evon Z. 1960. 'On the Concepts of Structure and Process in Cultural Anthropology', *American Anthropologist* 62(1): 18–33.

Vogt, Evon Z. 1964. 'Ancient Maya and Contemporary Tzotzil Cosmology: A Comment on Some Methodological Problems', *American Antiquity* 30(2): 192–5.

Vogt, Evon Z. 1965. 'Structural and Conceptual Replication in Zinacantan Culture', *American Anthropologist* 67(2): 342–53.

Vogt, Evon Z. 1969. *Zinacantan. A Maya Community in the Highlands of Chiapas.* Cambridge MA: Harvard University Press.

Vogt, Evon Z. [1990] 1970. *The Zinacantecos of Mexico: A Modern Maya Way of Life.* New York: Holt, Rinehart and Winston.

Wagley, Charles. 1949. *The Social and Religious Life of a Guatemalan Village.* Menasha WI: American Anthropological Association.

Wagner, Roy. [1975] 2016. *The Invention of Culture – Revised and Expanded Edition.* Chicago IL: University of Chicago Press.

Wagner, Roy. 2010. *Coyote Anthropology.* Lincoln NE: University of Nebraska Press.

Warman, Arturo et al. 1969. *De eso que llaman Antropología Mexicana.* México: Comité de Publicaciones de los Alumnos de la ENAH.

Watanabe, John M. 1992. *Maya Saints and Souls in a Changing World.* Austin TX: University of Texas Press.

Watanabe, John M., and Edward F. Fischer. 2004. *Pluralizing Ethnography: Comparison and Representation in Maya Cultures, Histories and Identities.* Santa Fe NM/Oxford: School of American Research Press/James Currey.

Weeks, John M., Frauke Sachse and Christian M. Prager. 2013. *Maya Daykeeping: Three Calendars from Highland Guatemala.* Boulder CO: University of Colorado Press.

West, Harry. 2005. *Kupilikula: Governance and the Invisible Realm in Mozambique.* Chicago IL: University of Chicago Press.

West, Harry. 2007. *Ethnographic Sorcery.* Chicago IL: University of Chicago Press.

Wilhelm Ludwig, Carl. 1922. *Between Two Continents: Notes from a Journey in Central America.* London: Eveleigh Nash and Grayson.

Wilson, Richard. 1995. Maya Resurgence in Guatemala: Q'eqchi Experiences. Norman OK: University of Oklahoma Press.

Wolf, Eric. 1957. 'Closed Corporate Peasant Communities in Mesoamerica and Central Java', *Southwestern Journal of Anthropology* 13(1): 1–18.

Woodward, Ralph Lee, Jr. 1993. *Rafael Carrera and the Emergence of the Republic of Guatemala, 1821–1871.* Athens GA: University of Georgia Press.

Wyld Ospina, Carlos [1929] 1967. *El autócrata: Ensayo Político Social.* Guatemala: Ministerio de Educación.

Zamora Corona, Alonso. 2019. 'El rostro de los días: sobre la corporalidad y las almas entre los mayas k'iche' de Santiago Momostenango', *Journal de la Société des Américanistes* 105(2): 107–32.

Zamora Corona, Alonso. 2020. 'Coyote Drums and Jaguar Altars: Ontologies of the Living and the Artificial among the K'iche' Maya', *Journal of Material Culture* 25(3): 1–24.

Zamora Corona, Alonso. 2021. 'The Face of the World: A Study on the Cosmological among the K'iche' Maya of Momostenango', PhD thesis, University College London.

Zamora Corona, Alonso. 2022. 'Towards a Complex Theory of Writing: The Case of Aztec and Mixtec Codices', *Signata: Annales des Semiotiques* 13: 350–83.

Index

www.ingramcontent.com/pod-product-compliance
Ingram Content Group UK Ltd.
Pitfield, Milton Keynes, MK11 3LW, UK
UKHW021836150926
8656IPUK00004B/19

9 781836 950264